Five-Minute Mathematics

Five-Minute Mathematics

Ehrhard Behrends

Translated by David Kramer

This edition is published by the American Mathematical Society
under license from Friedr. Vieweg & Sohn Verlag.

Originally published in the German language
by Friedr. Vieweg & Sohn Verlag, 65189 Wiesbaden, Germany,
as "Ehrhard Behrends: Fünf Minuten Mathematik.
1. Auflage (1st edition)".

Translated by David Kramer

2000 *Mathematics Subject Classification.* Primary 00A06, 00A08.

For additional information and updates on this book, visit
www.ams.org/bookpages/mbk-53

Library of Congress Cataloging-in-Publication Data

Behrends, Ehrhard, 1946–
[Fünf Minuten Mathematik. English]
Five-minute mathematics / Ehrhard Behrends ; translated by David Kramer.
p. cm.
Includes index.
ISBN 978-0-8218-4348-2 (alk. paper)
1. Mathematics—Popular works. 2. Mathematical recreations. I. Title.

QA93.B4413 2008
510—dc22 2007060594

∞ The paper used in this book is acid-free and falls within the guidelines
established to ensure permanence and durability.
Visit the AMS home page at http://www.ams.org/

10 9 8 7 6 5 4 3 2 1 13 12 11 10 09 08

Contents

Foreword: "Five-Minute Mathematics" in *Die Welt*

by Dr. Norbert Lossau, editor-in-chief for science at *Die Welt* and author of the column "Five-Minute Physics"

Most people do not harbor any particular affection for mathematics. They perceive numbers and formulas as difficult, confusing, abstract, irrelevant. And perhaps it is true that one needs a certain predisposition, analogous to what we call musicality, to develop a passionate interest in mathematics.

Nevertheless, I am convinced that many mathematical skeptics would readily take an interest in the queen of the sciences if only someone would build them a bridge into the fascinating realm of mathematics. Teachers could build such bridges by packaging mathematical lessons in suspense-filled stories from daily life. What would be the response, for example, if the discussion of abstract curves were motivated by a search for the optimal terms of a stock option? Or if one used geometry to calculate the amount of living space in a geometrically complicated dwelling and the number of rolls of wallpaper needed for redecorating? And when it comes to prime numbers, a story about cryptography and the challenges of breaking secret codes is certain to get the attention of many pupils.

The science of mathematics is central to our lives. It is to be found everywhere one looks: from the scanner at the checkout counter, the calculation of mortgage interest, and the PIN code on your debit card to computer tomography in medicine and the design of automobiles and airplanes. Mathematics makes possible space probes to distant planets and brings robots to life. It is the driving force behind technological advances and—if one allows oneself to be drawn into the subject—simply unbelievably fascinating.

Even if no bridge was laid down in far-off schooldays, there are still opportunities for grownups to approach the subject. For one thing, the coverage of science and technology in the media has improved enormously in recent years, although alas, such cannot be said for coverage of mathematics. Only a few newspapers and television outlets report regularly, or even sporadically, on topics related to mathematics even though there is much that is worth reporting. It would appear that many editors consider the subject of mathematics taboo.

Die Welt does not suffer from such fears and is not afraid, for example, of devoting a double-page spread to the number π (25 February 2006).

With the weekly column "Five-Minute Mathematics" from the pen of Professor Ehrhard Behrends, the newspaper has provided a stable editorial forum for the publication of one hundred columns on mathematical topics. From the large number of readers' responses, we know that the column has generated considerable interest. Mathematics has been communicated—packaged in motivational stories—concisely and succinctly, comprehensibly and competently. And wonder of wonders, the unpalatable subject of mathematics has suddenly developed a pleasing taste.

"Five-Minute Mathematics" deserves to reach readers beyond the subscribers to the newspaper *Die Welt*, and we are pleased that with this book, the publishing house Vieweg Verlag is making this series of one hundred columns available to a wider public.

Professor Behrends is a bridge-builder, offering a way across the moat haunted by the dragons of math anxiety into fortress mathematics. He has the ability to package mathematical content so skillfully that there are few traces remaining of arid abstraction. If the status of mathematics is finally gradually to rise in the public's perception, we need more writers like him, and of course more publications that will offer forums to these authors.

Translator's Note

Translating Ehrhard Behrends's hundred mathematical vignettes has been a great pleasure. It was also occasionally a challenge. The phrase "lost in translation" exists for a reason, since no translation can reproduce every nuance of the original. Fortunately, mathematics is a universal language, and I was not confronted with the types of issues faced by the translator of poetry or fiction. I was able to carry over most of the content intact. Some choices had to be made, however. Should I, for example, change references to the currency of the European Union, the euro, to dollars? I decided that the English-speaking audience could deal with a foreign currency, and so euros were retained. As of the moment of writing, one euro can be purchased for about US$1.42. On Monday, 12 May 2003, when the first of these articles appeared, the euro was trading at US$1.15.

I also retained geographical references to Germany and references to the German national lottery. On the other hand, for the English reader's benefit, kilograms and meters have been converted to pounds and feet, and degrees Celsius to degrees Fahrenheit. References to odds of obtaining particular hands in the card game skat have been reinterpreted in terms of poker.

That was the easy stuff. How was your translator to deal with the chapter title "Die Mädchenhandelsschule," in which the way that items in a mathematical expression are grouped is explained by analogy to natural language? A *Handelsschule* is a business school, and so a *Mädchenhandelsschule* is a business school for girls. Unless, that is, one misreads it as *(Mädchenhandels)schule*, in which case one has something along the lines of a school for trading in girls. The German language is not associative! Neither is the English language, and I sought to come up with equivalent expressions. Lest the reader entertain any doubt, the discussion of the 1958 hit song "Flying Purple People Eater" in Chapter 20 of this volume does not appear in the German original.[1]

No man is an island, and my work benefited from the collaboration of a number of people, whom I would like to thank here. Everyone at the American Mathematical Society with whom I worked was unstintingly helpful and friendly. I would like to thank Ed Dunne, who invited me to undertake this project, and Cristin Zanella, who kept me in the loop and answered all my queries. Thanks also to Barbara Beeton for her friendly and intelligent TEXnical support and to Arlene O'Sean for her careful copyediting.

I owe special thanks to two individuals who read the translation as it was being produced. One of them, Professor Ehrhard Behrends, is of course the author of these articles, and his careful reading made it possible for me to correct a number of typographical errors and clarify some fuzzy points.

My second reader was Christina Kramer, one of my several sisters, who as a professor not of mathematics but of Slavic linguistics brought to the book the analytical skills of the linguist and the intelligence of the "educated reader" along with the innocent eye of the mathematical neophyte. She regrets that she is perhaps one of those for whom Chapter 31, "I Hate Mathematics Because...," was written. Christina called my attention to a number of linguistic anomalies and corrected quite a few typographical errors. She also pointed out several places where an additional phrase or sentence could rescue the reader with a background similar to hers from total befuddlement.

---

[1] Is it a (purple people) eater or a purple (people eater)?

Finally, I would like to thank my dog, Orpheus, without whom this translation would have gotten done much sooner, but at the price of my not having taken nearly as many salubrious walks.

Preface to the English Edition

A few weeks after my book *Fünf Minuten Mathematik* went to press, I received a proposal from the American Mathematical Society to publish an English-language edition. I must say that I was pleased at the prospect, although it would mean quite a bit of work for me, since I would surely have to revise some of the chapters extensively, and one or two would have to be completely rewritten. After all, would English-language readers have any interest in the German lottery? Or be able to make heads or tails out of it even if they were? And were there enough math-haters out there to justify reprinting the chapter "I Hate Mathematics"? And how could anyone possibly translate the examples using colloquial German that I chose as illustrations of mathematical laws?

Having read the English translation, I see that I had no need for worry. The reason: the translator, David Kramer, to whom I would like to offer here my most heartfelt thanks. Of course he translated my German sentences into perfect English, which is no more than one would expect from a pro. But he has achieved much more, for through an intensive email correspondence between the two of us over several weeks, I have seen how the book has benefited as well from a number of amplifying remarks (from which future German editions will also profit).

David has also added a number of observations of particular interest to the English-speaking world, and if that weren't enough, he found a number of typographical errors in the German edition that had somehow eluded everyone else.

In my opinion, his masterpiece is the translation of Chapter 20. In second place are some of the chapter and section titles, such as "The Butterfly That Fluttered By" and "Both a Borrower and a Lender Be." You can't do much better than that.

I wish you, dear reader, much pleasure in perusing these "Five-Minute" morsels of mathematics, and I hope that those among you who may be skeptical about anything at all to do with mathematics will be disabused of the beliefs that everything interesting has already been discovered and that mathematics is nothing more than a bone-dry collection of facts and techniques.

Preface to the German Edition

During the years 2003 and 2004 there appeared the first and, so far, only regular column on mathematics in a newspaper read throughout Germany. "Fünf Minuten Mathematik" appeared every Monday in *Die Welt*, and the *Berliner Morgenpost* reprinted the column several weeks later.

By the end of two years, one hundred columns had been published, covering a wide variety of topics. Regular readers of the column obtained an overview of cryptography and coding theory, as well as insights into the fascination of prime numbers and the infinite, mathematics in the CD player and CAT scan, the notorious Monty Hall problem and other mysteries of probability theory, to name but a few of the subjects covered.

This book contains all one hundred articles. They have been carefully revised and expanded with explanatory texts, tables, and figures that have more than doubled the original length.

Everyone with an interest in learning more about aspects of contemporary mathematics that can be explained without assuming any specialized knowledge will find something of interest in these pages.

The author especially hopes to convince readers who were traumatized by school mathematics that the subject is not the boring, dry-as-dust subject that they remember, but a wellspring of fascination and excitement.

Acknowledgments

The American Mathematical Society gratefully acknowledges the kindness of the following institutions and individuals in granting the following permissions:

Elke Behrends

Graphics of the rice mountain in Chapter 6, the monkey in Chapter 10, and Hilbert's Hotel in Chapter 15.

Bertrand Russell Archives, McMaster University Library

Photograph of Bertrand Russell at the blackboard in Chapter 8.

Vagn Lundsgaard Hansen

Photograph of the bridge in Chapter 53.

Robin Wilson

Photograph of Andrew Wiles in Chapter 89 taken and owned by Robin Wilson.

The author wishes to express his gratitude to the American Mathematical Society's editorial and production departments for the careful preparation of this book.

Introduction

The story of this book begins on 25 January 2002, when the German Mathematical Society decided to hold a dinner to bring together the society's officers and a group of journalists. The agenda was a conversation about the image of mathematics in the world at large. One of the participants was Dr. Norbert Lossau, science editor for the newspaper *Die Welt*, with whom I met again several months later. Out of these conversations arose the idea of a regular column on mathematics.

I put together an extensive proposal, in which about 150 possible topics were sketched. My suggestion of "Five-Minute Mathematics" as the title of the column was accepted, the graphic designers came up with a logo, and the first column appeared in the Monday, 12 May 2003, edition of *Die Welt*. And so it went week after week, with the rhythm being broken only when Monday was a holiday and the newspaper did not appear. After two years and one hundred columns, "Five-Minute Mathematics" yielded to another column.

In my selection of topics I have attempted to think particularly of readers who left school long ago and perhaps have no concrete memory traces of the subject yet would like to learn something about mathematics. Should the quadratic formula and curve sketching be the limit of what is worth learning about mathematics? Where is mathematics to be found in the "real world"?

In two years I was able to cover a wide spectrum of topics, as can be seen from a perusal of the table of contents. There is the contemporary and there is the classical; there are hors d'oeuvres and main courses. And in many places the reader will learn how mathematics penetrates our lives, whether in the lottery, cryptography, computer-aided tomography (CAT), and the evaluation of securities options.

Even before the final column appeared, I received a proposal from the publisher Vieweg to collect the columns in a book. There were many good reasons to begin at once. First, many regular readers of the column had inquired about such a book. Second, a newspaper column is confined to a fixed size, so that every column had to have the same length, regardless of the topic.[1] For some of the topics, the space limitation meant that important information had to be omitted, leaving the author with a guilty conscience. Therefore I am pleased that the book format allows such limitations to be overcome. And finally, the luxury of space in a book means that the word can be supplemented by the image: photographs, drawings, graphs, tables....

In writing the column there were three aspects that I considered important:

Mathematics is useful: It should be made clear why our technologically based world could not function without mathematics. A label reading "mathematics inside" could be placed on many a product.

Mathematics is fascinating: Aside from its utility, mathematics offers a very special intellectual appeal. The irrepressible compulsion to see the solution of a problem through to the end can release enormous amounts of energy.

Without mathematics one cannot understand the world: According to Galileo, "The book of nature is written in the language of mathematics." At his time, that was no more than a vision. Today it is known that mathematics is the bridge that leads us across the

[1] At least that is what the author was told. Every now and then, the exigencies of the page layout required that the column be trimmed a bit.

unknown into realms that lie beyond the limits of human perception. Without mathematics, it would be impossible, in Goethe's words, "to know what holds the world together from the inside."

I would like to thank Dr. Lossau for allowing me for two years to present mathematical topics to readers of *Die Welt*. I retain wonderful memories of our collaboration.

I wish also to thank Elke Behrends for many photographs, particularly the photomontages appearing in Chapters 6, 10, and 15. I am also grateful to colleagues Vagn Hansen (Copenhagen) and Robin Wilson (Oxford) for the images that they provided (Chapters 53 and 89). Finally, I would like to thank Tina Scherer and Albrecht Weis for their extirpation of so many typographical errors during proofreading so that you, dear reader, will not have to be annoyed by them.

Chapter 1

You Can't Beat the Odds

Suppose, for the sake of argument, that you live in a large city such as Berlin or Hamburg. You are seated on a bus; a passenger departs, leaving behind an umbrella. You take the umbrella, with the idea that when you get home, you will pick up the phone and dial seven random numbers in the hope of reaching the owner of the umbrella.

This is, of course, a made-up story, and such a plan in real life would be ridiculed as hopelessly naive. But don't laugh too quickly, because many of your fellow citizens have the hope every Saturday evening of having chosen the correct lottery numbers, the probability of which is 1 in 13,983,816. Such odds are worse than those of locating the owner of the umbrella according to the plan described above, since there are "only" ten million random sequences of seven digits.

Many lottery players imagine that they can outwit chance by choosing numbers that have not appeared frequently in the past. Such a strategy is wholly without merit, for chance has no memory. Even if, say, the number 13 hasn't been drawn in a long time, in today's drawing it has exactly the same probability of being chosen as any of the other numbers. Other lottery players swear by their own cleverly devised systems for beating the odds, but all such attempts are nothing but wasted effort, for it has been many decades since it was proven that there is no system that can fool chance.

Let us close with a bit of advice: In fact, there is some positive action that a lottery player can take, and that is to choose a combination of numbers that is unlikely to be chosen by many other players. Then if, by some small chance, one wins, one is less likely to have to share the prize with a large number of other winners. That, however, is easier said than done. On one recent occasion, many lottery winners saw their dreams of millions greatly reduced when it turned out that the winning numbers, which formed a cross on the selection card, had been chosen by a surprisingly large number of people.

In the end, however, mathematics has nothing to say about the sweet feeling of expectation that inspires plans about all that one might do with one's fabulous winnings. I wish you good luck!

And Why Exactly 13,983,816?

How does a mathematician arrive at the precise number 13,983,816 of possible selections of lottery entries? Choose two numbers, let's call them n and k, and let us assume that n is larger than k. How many different k-element subsets are contained in a set of n objects?

While this may seem an abstract mathematical question, it concerns us directly in the question of the lottery. A lottery entry is, after all, nothing other than a selection of six numbers from among the numbers 1 through 49. So in this case we are dealing with the numbers $n = 49$ and $k = 6$.

We can easily find similar examples from our common experience:

- For $n = 52$ and $k = 5$ we are asking about the number of possible hands in poker.
- If at the close of a committee meeting each of the fourteen members shakes hands in parting with each of the others, how many handshakes are involved? Here we have the case $n = 14$, $k = 2$.

And now back to the general problem. The formula that we are seeking is a fraction with numerator $n \cdot (n-1) \cdots (n-k+1)$ and denominator $1 \cdot 2 \cdots k$. The numerator may look a bit frightening to the uninitiated, but it is simply the product of the k whole numbers counting down by 1 from n. (Those interested in learning about where this formula comes from will find an introduction in Chapter 29.)

Here are a few additional examples:

- For the lottery problem, we must divide $49 \cdot 48 \cdot 47 \cdot 46 \cdot 45 \cdot 44$ by $1 \cdot 2 \cdot 3 \cdot 4 \cdot 5 \cdot 6$. That is where the number 13,983,816 comes from.
- For the poker problem, the quotient is $52 \cdot 51 \cdot 50 \cdot 49 \cdot 48$ divided by $1 \cdot 2 \cdot 3 \cdot 4 \cdot 5$, which leads to 2,598,960 different poker hands. Note that since only four of these hands are royal flushes, the probability of being dealt such a hand is $4/2{,}598{,}960 = 1/649{,}740$, or about three royal flushes out of every two million hands dealt.
- You can solve the handshake problem in your head: $14 \cdot 13$ divided by $1 \cdot 2$ is 91.

A Four-Mile-High Stack of Cards

The idea of randomly selecting the telephone number of an unknown person as an aid in coming to grips with the tiny probability of winning the lottery is not the only possible example. Here is another one.[1]

We begin with the observation that a deck of cards placed on the table is about an inch thick. It would take about 270,000 decks of

[1] Yet another way of picturing this small probability is offered in Chapter 83.

cards to assemble a stack of 13,983,816 cards. Since 270,000 inches is 22,500 feet, which is about 4.25 miles, a stack of 13,983,816 cards would be over four miles high. Suppose now that just one of those cards has a check mark on it. The odds of selecting that card at random on a single try from the four-mile-high stack of cards are about the same as the odds of choosing the correct six lottery numbers.

Chapter 2

Magical Mathematics: The Integers

I would like to introduce to you a little guessing game. Choose a three-digit number and write it twice in succession. For example, if you chose 761, then you should write down 761,761. The game begins by dividing your six-digit number by 7. The remainder, that is, whatever is left after the division, is your lucky number. This will be one of the numbers $0, 1, 2, 3, 4, 5, 6$, since these are the only possible remainders on division by 7. Now write your number and the remainder on a postcard and send it to the editor of this newspaper (*Die Welt*). By return post you will receive as many 100-euro notes as indicated by your lucky number.

$$
\begin{array}{r}
108823 \\
7\,\overline{)\,761761} \\
\underline{7} \\
06 \\
\underline{0} \\
61 \\
\underline{56} \\
57 \\
\underline{56} \\
16 \\
\underline{14} \\
21 \\
\underline{21} \\
0
\end{array}
$$

If you are unfortunate enough to have ended up with zero as your lucky number, you are in good company, since the same fate will have befallen all of your fellow readers. (If such were not the case, the publisher would never have agreed to print the newspaper article.)

The reason for this phenomenon rests in a well-hidden property of the set of whole numbers, or integers. Namely, placing a three-digit number next to itself is equivalent to multiplying it by 1,001, and since 1,001 is divisible by 7, the six-digit number will be divisible by 7 as well.

This idea can be packaged as a little magic trick for one's private use; one can replace the promise of 100-euro notes by predicting the remainder.

Indeed, it happens frequently that a mathematical fact somehow finds its way into a magician's hat. One simply has to find mathematical results that contradict everyday experience and that also have their basis hidden in the depths of some theory.

Here is a piece of advice: Magic is like perfume: the packaging is at least as important as the contents. No one should be suggesting that the chosen three-digit number is to be multiplied by 1,001; such a multiplication is equivalent to writing the number twice in succession, but then the whole trick would fall flat. Those looking for a variant from dividing by 7 can substitute 11 or 13, since 1,001 has these numbers as factors as well. It will just make the calculation of the remainder a bit more difficult.

Advanced Variants: 1,001, 100,001, ...

Is there a reason that we have to write down precisely a *three*-digit number? Could we achieve a similar result with two or four digits?

Let us consider a two-digit number n, written in the form xy. If we write the number twice in succession, then we obtain the four-digit number $xyxy$, which is equivalent to multiplying the original number by 101. But 101 is a prime number, and so the divisors of $xyxy$ are the divisors of xy together with 101. Since in performing this magic trick we know nothing about the number xy, we can say only that there will be zero remainder on division by 101. But asking for division by 101 gives the trick away, or at least strongly suggests what is at work, and furthermore, dividing by 101 may be too difficult for your friends and acquaintances. We conclude, then, that starting with a two-digit number is not such a good idea.

With four-digit numbers we are dealing with multiplication by 10,001. This number is not prime, since $10{,}001 = 73 \cdot 137$, with both factors being prime. Therefore, if you write down a four-digit number twice to form an eight-digit number, it is guaranteed that it is divisible by 73 and 137. But who is eager to divide by 73?

Since the number 100,001 has only the prime divisors 11 and 9,091, both inconvenient divisors, five-digit numbers are not optimal for our magic trick. And so it goes. We again find small divisors with 1,000,000,001 (it is divisible by 7). But do we really want to begin our magic act with, "choose a nine-digit number and write it down twice to form an eighteen-digit number"? My recommendation is that you stick with the original trick.

Here is a table of prime factors for the first several numbers of the form $10\ldots01$:

Number	**Prime Decomposition**
101	101
1,001	$7 \cdot 11 \cdot 13$
10,001	$73 \cdot 137$
100,001	$11 \cdot 9{,}091$
1,000,001	$101 \cdot 9{,}901$
10,000,001	$11 \cdot 909{,}091$
100,000,001	$17 \cdot 5{,}882{,}353$
1,000,000,001	$7 \cdot 11 \cdot 13 \cdot 19 \cdot 52{,}579$
10,000,000,001	$101 \cdot 3{,}541 \cdot 27{,}961$
100,000,000,001	$11 \cdot 11 \cdot 23 \cdot 4{,}093 \cdot 8{,}779$
1,000,000,000,001	$73 \cdot 137 \cdot 99{,}990{,}001$

Those wishing more information about the relations between mathematics and magic could do worse than to consult Martin Gardner's book *Mathematics, Magic, and Mystery.* Additional magic tricks with a mathematical basis will be presented in Chapters 24 and 86.

Chapter 3

How Old Is the Captain?

Mathematics is considered—and rightly so—a particularly exact science. Its strictly logical construction has served as a model for many other fields in the natural sciences and humanities. A famous example of this is Isaac Newton's magnum opus, *Philosophiae Naturalis Principia Mathematica*. It begins with fundamental definitions and axioms about the world (what is force? what is mass? what are the fundamental laws of motion?), and out of this is derived—in a strictly deductive manner—a model of the world that revolutionized science.

After Newton, there arose a belief in the nature of progress that to us today seems rather naive: all phenomena should be reduced to the simplest possible mechanical model. Many of our fellow citizens still have the tendency to give particular credence to assertions that are couched in mathematical terminology, perhaps even embellished with a mathematical formula. But a good dose of skepticism is frequently in order, for usable results can be expected only when they are based on clear underlying concepts. Thus we will certainly all agree on a definition of "velocity," while "perceived temperature," on the other hand, is an altogether subjective matter. And therefore, the wind-chill formula, for example, is something that one may consider, depending on one's taste, as either amusing or annoying.

In this regard, one would do well to keep in mind the natural limits of mathematics. No matter how much intelligence is brought to bear on a topic, no valid result can be derived from insufficient information. Sometimes, the "result" is hidden so impossibly in the problem that the whole thing is to be taken as a joke: "A ship is 45 meters long. How old is the captain?"

In this form, it is clear to all that such questions are nonsensical. Nevertheless, one frequently encounters questions of the form, "What is the probability that Germany will become world champion?" And how is one to evaluate the odds of winning some soap company's sweepstakes when no one knows how many prizes are to be awarded and how many contestants there are?

Wind Chill and Related Matters

One formula that is used for perceived temperature due to wind chill is

$$T_{\mathrm{wc}} = 35.74 + 0.6215T - 35.75v^{0.16} + 0.4275Tv^{0.16},$$

where T is the actual temperature in degrees Fahrenheit and v is the wind speed in miles per hour. For example, if $T = 32$ and $v = 12$, then T_{wc} is 22.8°F.

This wind-chill formula is a nice example of falsely conceived exactitude. Everyone agrees that it feels colder than it actually is when a strong wind is blowing. But it would be difficult to find two individuals for whom the "perceived temperature" at, say, 22 degrees Fahrenheit and a wind speed of 7 miles per hour was the same to four decimal places of accuracy. The temperature that is "felt" depends on the individual's constitution and clothing, as well as a host of other factors.

But the wind-chill calculators act as though they could calculate perceived temperature exactly. They even present a formula somehow cobbled together from the various parameters, and yet with an accuracy to four decimal places. To be sure, one expects monotonicity: when the wind is stronger, the temperature feels colder. But all in all, we would be better off with a very rough table of values, because the

$$h = Q(12 + 3s/8)$$

Figure 1. A mathematical attempt at humor. The formula supposedly gives the optimal height of a woman's high heels as a function of the number of drinks consumed.

formula leads to a wholly unjustified impression that we are dealing here with an exact science.

In the meanwhile, a host of imitators have appeared on the scene. For example, there have been reports of formulas for the height of stiletto heels (see Figure 1) and the degree of tension produced by a suspense novel. Such "scientific" attempts are frequently to be found in the "Features" sections of newspapers. And thus one can marvel, while eating one's breakfast toast and reading the morning paper, at how mathematics has been pressed into service as a source of humor.

Chapter 4

Vertiginously Large Prime Numbers

The simplest numbers are surely the so-called natural numbers, the ones we use to count: $1, 2, 3, \ldots$. Some of these numbers enjoy the special property that they cannot be written as a product of smaller numbers. Such is the case for 2, 3, and 5, but also for 101 and 1,234,271. Such numbers are said to be *prime*, and they have exercised particular fascination since the earliest beginnings of mathematics.

How big do the prime numbers get? Over two thousand years ago, Euclid gave a famous proof of the fact that there are infinitely many prime numbers, and therefore prime numbers of arbitrarily large size. The idea is the following: Euclid describes a sort of machine into which one deposits some prime numbers, and out comes a prime number that is different from all the prime numbers that were put in. It then follows that there cannot be a finite number of prime numbers.

The consequences of this fact are remarkable, some so remarkable that they can induce in certain individuals a feeling of vertigo. For example, Euclid's result guarantees that there exists a prime number so large that to print it would take more ink than has been produced in the history of the world; we shall, of course, never see such a monster in the flesh. The largest prime number that has been positively identified as such (in the year 2006) has almost ten million digits.

(To get an idea of the size of this number, if you wanted to publish a book in which this record-holding prime was to be printed, it would require over two thousand pages.) Large prime numbers are of use in cryptography, but for practical applications one can use "small" primes of only a few hundred digits.

One of the major tasks of the field of mathematics known as number theory is to discover new secrets about the primes. The great mathematician Carl Friedrich Gauss called number theory the "queen of mathematics."

The Prime Number Machine

Here we give a functional definition of Euclid's prime number machine. Suppose we are given n prime numbers, which we shall name $p_1, p_2, \ldots, p_n$. If such a description seems too abstract, then just keep in mind the four prime numbers $7, 11, 13, 29$, in which case $n = 4$ and $p_1 = 7$, $p_2 = 11$, $p_3 = 13$, and $p_4 = 29$.

We now form the product of these primes and add 1. We will call the result m. Thus

$$m = p_1 \cdot p_2 \cdots p_n.$$

In our special example, we have $m = 7 \cdot 11 \cdot 13 \cdot 29 + 1 = 29{,}030$.

Every number, and therefore m, has at least one prime divisor. Let us call such a divisor p. Observe that p must be different from all of p_1, p_2, $\ldots$, p_n, since if m is divided by one of these numbers, the remainder is 1. (In our example, we could choose $p = 5$, which is a prime divisor of 29,030; and indeed, the number 5 is not to be found among $7, 11, 13, 29$.)

Putting it all together, we see that for an arbitrary collection of prime numbers $p_1, p_2, \ldots, p_n$, a new prime number is produced that was not part of the input. Thus it cannot be the case that the collection of prime numbers is finite, because any finite set of primes fed into the machine produces yet another prime.

Figure 1 shows some additional examples in which now the output is *all* prime divisors of $p_1 \cdot p_2 \cdots p_n + 1$. Note particularly the second and third examples: the prime numbers that are input do not have to be distinct.

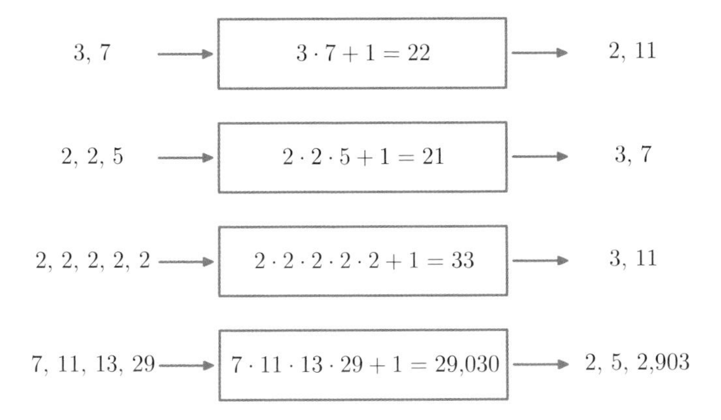

Figure 1. Euclid's prime number machine in action.

Does Euclid's machine generate *all* prime numbers? What we mean by this is the following: We take it as known that 2 is a prime number. We feed it into the machine, and out comes the prime number 3 (the "product" of a set of a single number is the number itself, so the output is $2 + 1$). Now we can feed the machine with 2 and 3, which yields 7, and so we can now work with 2, 3, and 7. These three primes are possible inputs, and not all of them have to be input at once, and one or more of them may be used more than once. The question is, does every prime number appear as output from Euclid's machine?

The answer is yes, since for every prime number p, $p - 1$ is the product of certain (not necessarily distinct) primes $p_1, \ldots, p_r$. Therefore, with input $p_1, \ldots, p_r$, the prime p will be output, since $p_1 \cdot p_2 \cdots p_r + 1 = p$. This argument can be used to prove, using mathematical induction, the assertion that all prime numbers smaller than the number n are output by Euclid's machine, where n is an arbitrarily large number.

Chapter 5

Loss Plus Loss Equals Win

Mathematics, and probability theory in particular, is chock full of surprising phenomena. When a result stands in stark contrast to general expectation, it is called a *paradox*. Not long ago, the Spanish physicist Juan Parrondo enriched the menagerie of such paradoxes with a new example.

Consider two games of chance in which the player loses on average a small amount to the house. One has to pay an ante to enter the game, and then one wins or loses one euro in each round with probability $\frac{1}{2}$. For the second game, the odds of winning depend on the prior course of the game. There are more- and less-favorable rounds, but on average, the odds are 50–50.

And now for the surprise: If before each round one tosses a coin to determine whether to play game 1 or game 2, then the player has a winning strategy. If the house is willing to let the player continue for a sufficiently long time, one can become arbitrarily wealthy. After

Parrondo's discovery, one could read in various places that now there was a mathematical theory for all possible situations in which an apparent losing proposition ended in a win. Everyone has had such experiences. For example, in a game of chess you can sacrifice almost every piece and yet emerge the victor.

Of course, there is no such theory. However, it is of interest that mathematical results that find their way from the ivory tower of academia to the daily newspapers almost always lead to great expectations that can never be fulfilled. Such was the case, as many will recall, with fractals and chaos theory. Nevertheless, a number of interesting applications of Parrondo's paradox have been found. For example, it explains how a microorganism can alternate between two chemical reactions to swim against the current.

The Precise Rules for Game 2

The rules for the Parrondo's first game have already been described. Those for the second game are somewhat more complicated:

> If the total amount that the player has won thus far is divisible by 3, then the odds are unfavorable: the player loses one euro with probability 9/10 and wins a euro with probability 1/10.[1]
>
> The situation improves when the amount of winnings is not divisible by 3. Then the player wins with probability 3/4 and loses with probability 1/4.

Thus there are favorable and unfavorable situations for the player depending on the divisibility by 3 of the amount won. It can be shown that the game is perfectly fair. However, because of the ante, it is a losing game in the long run.

[1] For example, one card can be drawn at random from a pack of ten. On nine cards is written, "you have just lost one euro!" and on the tenth appears, "you have won one euro!"

A Paradox!

Paradoxes exist in many branches of mathematics. They are to be expected when there are phenomena that are inaccessible to our direct experience: very large or very small numbers, infinite sets, etc.[2]

It is a bit surprising that paradoxes appear so frequently in probability theory, since we have obtained in the course of human evolution a good feel for many aspects of chance. For example, we can gauge the mood of our interlocutor quite well based on facial expression alone, and we can estimate simple risks quite well.

A well-known paradox is one involving birthdays, which is described in Chapter 11. Another well-known example is the *permutation paradox*: A man writes ten letters and addresses ten accompanying envelopes. He puts the letters in the envelopes, but the envelopes are chosen at random. Will at least one letter end up in the correct envelope? A naive assessment would conclude that such a probability is extremely small. However, probability theory tells us that the likelihood is a full 63%. Try it! (In the guise of choosing partners in a game, this paradox appears in Chapter 29.)

[2]Some paradoxes of the infinite can be found in Chapters 15 and 70.

Chapter 6

When It Comes to Large Numbers, Intuition Fails

The course of human evolution has ill prepared us for discoveries in physics and mathematics. Such things are of only marginal importance in reproduction and survival. Of interest are only such items as average velocity, lengths that are neither very large nor very small, and numbers of relatively small size. Thus just as it is difficult to understand the current view of the nature of the universe, where bizarre phenomena occur at very high velocities, so there is an almost inbuilt inability to grasp certain mathematical truths.

For example, let us talk about large numbers. In physics, there is at least the possibility of representing distances that go beyond our experience and intuition in terms of a suitable scale of measurement. Thus, for example, one can represent the solar system as a miniature model in which the sun is shrunk, say, to the size of an orange. With mathematics such opportunities of representation are fewer, and our ability to imagine what is going on is quickly left in the dust.

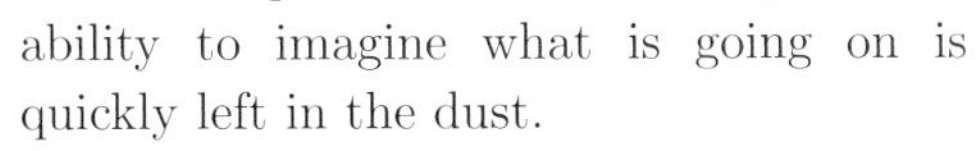

One thing that is particularly difficult to grasp is exponential growth. Many people have heard the parable of the grain of rice: If a grain of rice is placed on the first square of a chessboard, and twice that number on the second square, and twice *that* number on the third, and so on, then the numbers of grains of rice on the successive squares are $1, 2, 4, 8, \ldots, 2^{63}$. After these 64 steps, the number of grains is so large that it is far, far greater than the world's annual production of rice. To be sure, rice on a chessboard is not an everyday occurrence. Something analogous that is more in line with our experience is the phenomenon of chain letters. Say you receive a letter that already has been making the rounds, and you are supposed to send copies of the letter, with your name and address appended, to ten of your friends, who are to do the same in turn. Everyone whose name is more than five generations old on the letter is to be sent a picture postcard (or 100 euros, or whatever). This seems like a great idea, and from a naive point of view, there is profit (in postcards or euros) to be made. After all, one sends a single postcard to keep the system alive, and after a while, you receive a bushel basket of mail. (Surely a bushel basket would not suffice: if all the players do as they are supposed to, you can figure on over 100,000 cards.) However, such a game usually falls apart in the early stages, because too many people are sent too many letters by too many friends with a request to send ten letters.

Mathematicians have a particular respect for exponential growth. Problems whose degree of difficulty grows exponentially with the size of the input are considered especially difficult. Thus, for example, one attempts to show that the problem of breaking an encryption procedure is of an exponential nature.

Exponential Growth I: The Flood of Rice

How many grains of rice are actually involved in the rice parable? We need to sum $1 + 2 + 4 + \cdots + 2^{63}$. Such sums are easy to evaluate using the formula for a *geometric progression*:

$$1 + q + q^2 + \cdots + q^n = \frac{q^{n+1} - 1}{q - 1} \quad \text{for } q \neq 1 \text{ and } n = 1, 2, \ldots .$$

In our case, we obtain

$$\frac{2^{64}-1}{2-1} = 18{,}446{,}744{,}073{,}709{,}551{,}615 \approx 18 \cdot 10^{18}.$$

That is a lot of rice!

We have no intuition for such numbers. Indeed, even the fourteen million different combinations in the lottery leave our heads spinning. Let us at least try to get a handle on this amount of rice. A grain of rice is, after all, roughly a cylinder of diameter 1/20 of an inch and length about 1/3 inch. Thus about 1,200 grains of rice should fit in a cubic inch.[1]

Now we can do the math. If 1,200 grains fit in a cubic inch, then one would need $1{,}200 \cdot 12^3 \approx 2{,}000{,}000$ for a cubic foot, and $1{,}200 \cdot 12^3 \cdot 5{,}280^3 = 305{,}229{,}673{,}267{,}200{,}000 \approx 3 \cdot 10^{17}$ for a cubic mile. We should therefore divide the number of grains of rice by $3 \cdot 10^{17}$ to obtain the size of our rice pile in cubic miles: it turns out that one would need about 60 cubic miles.

To make such a number even more accessible, consider the following. Given that Germany has an area of 138,000 square miles, the amount of rice necessary to satisfy the requirements of the chessboard can be reformulated as follows: the rice would cause Germany to disappear under more than two feet of rice.

You don't believe it? I didn't believe it either, so I decided to try for myself. The result is pictured in Figures 1, 2, and 3.

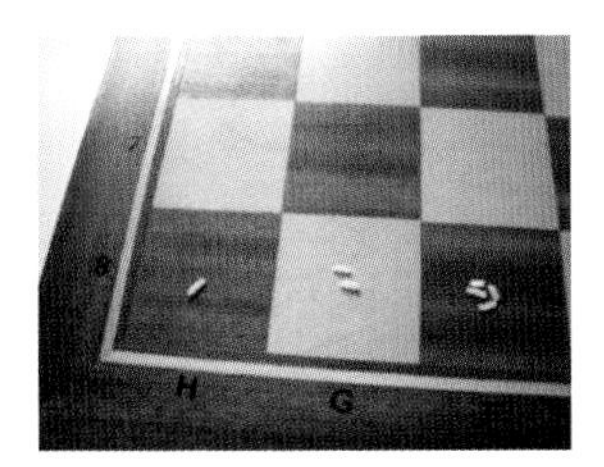

Figure 1. It began innocently enough...

[1] Even more if one could pack them optimally; here they lie every which way.

Figure 2. ...but then it went faster than I had expected...

Figure 3. ...and finally I had to give up.

Exponential Growth II: How Many Times Can You Fold a Sheet of Paper?

Before you read any further, answer the following question: How often do you think a piece of paper can be folded repeatedly in half? Most people guess wrong, giving much too high a number.

In folding, there are two aspects to consider. First, the thickness of the folded paper grows exponentially, doubling after each fold. After five folds, the folded paper is 32 pages thick, since $2 \cdot 2 \cdot 2 \cdot 2 \cdot 2 = 32$. That is about half an inch, and if you could manage another five folds, you would have about 16 inches.

But that is impossible: when several layers are arranged one above the other and have reached a thickness d, then the situation is different for the upper—that is, the inner after folding—layer than for the lower. Namely, the lower layer must be stretched, indeed enough to make a semicircle of radius r. The circumference of a circle is $2\pi r$, and so we are dealing here with the length πr. For example, if five folds represent half an inch thickness, then the lowest layer must compensate by being stretched by $\pi/2 \approx 1.6$ inches.

After a few more folds, the stretching has reached its limit. Experience shows that the limit is eight folds. (A Berlin radio station wanted to verify this, and on 12 September 2005 a public paper-folding was made with a sheet of paper of dimensions 33 by 49 feet. Even in this case, eight folds was the limit.)

Chapter 7

The Key for Encryption Is in the Telephone Book

It is a dream that goes back centuries, to find a method of sending secret messages in such a way that they really stay secret. Under the name "cryptography," the realization of this dream has become a branch of mathematics that today has a large number of practitioners.

It is interesting that the development of cryptographic methods has caused certain mathematical areas of specialization to come down from their ivory tower. Consider number theory, a venerable branch of mathematics that investigates the familiar numbers $1, 2, 3, \ldots$. In the past few decades it has suddenly become a matter of great commercial importance to know as much as possible about the prime numbers, since new results have much to say about the security of encrypted data transfer.

Cryptography has always been a reliable source of spectacular surprises. It began when it was discovered that it was no longer necessary to keep the information used in encryption and decryption a secret. Under the name *public key cryptography*, this idea revolutionized the discipline. Now security depended on a very special problem involving prime numbers: whoever is able to figure out the factors of a large number that is the product of two prime numbers will be able to read encrypted messages. Now, it is easy to see that, say, 35 is the product of the two prime numbers 7 and 5, but things become more difficult with the number 49,402,601 (which is the product of the primes 33,223 and 1,487). However, in serious cryptography, the numbers involved have hundreds of digits. It is generally believed that there is no procedure that can determine the factors of such large numbers rapidly enough to break a coded message in practical applications. Thus it came as a shock when it was demonstrated a few years ago that such factoring could be easily accomplished by a quantum computer, if such a thing ever were invented. In the meanwhile, cryptographers can rest easy, though they would feel much better if it could be proved conclusively that the systems in use today are indeed secure. Despite great efforts, that has not yet been achieved.

Random Keys Are Secure!

A more complete description of the relationship between cryptography and prime numbers appears in Chapter 23.

Even without prime numbers it is possible to come up with absolutely secure encryption procedures, provided that one is willing to put up with a small defect. The best-known of these goes like this: Toss a coin a large number of times, say 10,000, and record the result as a sequence of zeros and ones. (If you are too lazy to do the tossing yourself, you can have a computer do it virtually.) The sequence might begin something like this:

$$0010111011011100000\ldots.$$

This sequence will now be used to encrypt a message, which, for the sake of simplicity, we will assume is itself a sequence of zeros and ones.[1]

Let us assume, then, that the message that we wish to encode goes like this:

$$10111001100000011000\ldots.$$

For the encryption, we proceed as follows: First write down the random sequence and the message one above the other:

$$00101111011011100000\ldots,$$

$$10111001100000011000\ldots.$$

Whenever two of the same symbol lie one above each other (two zeros or two ones), we record a zero, and otherwise, a one. In our case, the result looks like this:

$$10010110111011111000\ldots.$$

This encoded message can now be sent without fear that anyone lacking the secret random sequence can make head or tail of it. For a recipient in possession of the secret key, decoding is simple. If, for example, the first symbol in the key is a zero, and the first symbol in the encoded message is a one, then the plain-text message must begin with a one (a zero would have resulted in a zero).

This procedure is absolutely secure. That is because all $2^{10,000}$ possible encrypted messages of length 1,000 have equal probability of being generated by the procedure. Unfortunately, this encryption method suffers from two serious defects. The first is that the recipient (as well as the sender) must be in possession of the secret key, and any transmission channel is subject to attack. A second drawback is that the key can be used only once. If it is used for a number of encryptions, it can be cracked using frequency analysis.

The mathematical public key methods do not suffer from these defects, and for that reason they are in wide use today.

---

[1]This could be accomplished, for example, by encoding the letters of the alphabet and the most important special characters as sequences of five zeros and ones: A = 00000, B = 00001, etc. Since $2^5 = 32$, one could encode 32 characters in this way.

Cryptography: A Secret Science

Cryptography is a branch of mathematics in which many discoveries have not been made public. An important area of current research deals with the question of how one can recover the factors of a number that is a product of primes. On this question turns the security of many encryption algorithms.

Surprisingly, in certain cases, factorization is a snap. But since it is not known publicly in which special cases current research has shown factorization to be easy, there is always a bit of uncertainty for those using large prime numbers for encryption.

An example should help in clarifying this issue. The idea goes back to Descartes. Suppose we have found a large prime number p. Beginning with p, we search among the next numbers for a further prime number q. Thus $q = p + k$, for some "small" integer k. As an illustration, consider the case $p = 23{,}421{,}113$, $q = 23{,}421{,}131$. Thus we have $k = 18$.

In actual applications, such numbers have several hundred digits, but Descartes's idea is already meaningful in the order of magnitude of our example. We now compute the product $n = p \cdot q$ and obtain $n = 548{,}548{,}955{,}738{,}803$. Is it possible to reconstruct p and q given n? If one guesses that one is dealing with $q = p + k$ with a not-too-large k, one might just hit pay dirt. Here is the idea.

It is clear that k must be an even number, since both p and q are odd. We write, then, $k = 2 \cdot \ell$. The number $p + \ell$, which lies halfway between p and q, will play an important role as the story unfolds. Let us call this number r. We then have $p = r - \ell$ and $q = r + \ell$. It follows that $n = (r - \ell) \cdot (r + \ell) = r^2 - \ell^2$, and therefore $n + \ell^2 = r^2$. We see, then, that n has the property that if a small perfect square is added to it, another perfect square results. This observation leads to the following strategy:

- Keep adding to n the perfect squares $\ell^2 = 1^2, 2^2, 3^2, \ldots$ and check each time whether $n + \ell^2$ is a perfect square. A computer can do this for you with no difficulty.

- When you get a positive result, write $n + \ell^2$ exactly as r^2.

- The desired factors p and q are given by $p = r - \ell$ and $q = r + \ell$.

In our concrete example, we are to investigate whether one of the numbers

$$548{,}548{,}955{,}738{,}803 + 1,$$
$$548{,}548{,}955{,}738{,}803 + 4,$$
$$548{,}548{,}955{,}738{,}803 + 9,$$

etc., is a perfect square. On the ninth attempt—that is, after a few milliseconds of computer time—we have success:

$$548{,}548{,}955{,}738{,}803 + 9^2 = 23{,}421{,}122^2.$$

We have merely to add 9 to 23,421,122 and then subtract 9 from 23,421,122 to obtain the two factors.

Chapter 8

The Village Barber Who Shaves Himself

There are not many German mathematicians who are known outside the borders of mathematics. Georg Cantor (see Figure 1), the founder of set theory, surely belongs to their number. Why is set theory important? Why does one speak of a "paradise of set theory" that is a sine qua non for mathematics?[1] The reason is that this branch of mathematics has made it possible to set mathematics on a strong deductive footing.

Figure 1. Georg Cantor and Bertrand Russell

[1]Thus was it formulated by the great mathematician David Hilbert.

Seen naively, set theory is quite harmless. One simply collects some objects of interest into a new object, called a set. Such set formation occurs in everyday life. It is clear, for example, that certain countries constitute the set "European Union" or that certain departments constitute the set "federal government." However, there is trouble when one allows the creation of new objects to proceed unchecked. Then the result can be nonsense, as was discovered a century ago by the British mathematician and philosopher Bertrand Russell (shown in Figure 1). His argument rests on a logical paradox that has been making the rounds since ancient times: when a statement is allowed to refer to itself, a logical breakdown is imminent.

A well-known dressing up of this paradox takes the form of a village barber, a man who has specialized in shaving precisely those men who don't shave themselves. But what about the barber? Does he shave himself? He can't, because he shaves only those who don't shave themselves. But if he doesn't shave himself, then he is one of his customers, and so he is in the group of those who don't shave themselves. You can twist it and turn it as you like; the question defies a logical answer.

After the shock of Russell's paradox, set theory recovered in part by not allowing self-referential sets, and today it is the undisputed foundation of mathematics.

Set Theory in Kindergarten

Older readers will recall that around 1960 there was a bull market in set theory. The cause was the sputnik surprise: in 1957, the Soviet Union launched the first artificial Earth satellite. The West responded with massive efforts to improve education in all disciplines, from kindergarten to university. Unfortunately, those involved in educational policy allowed themselves to be convinced that a basic mastery of set theory was necessary for understanding mathematics. As a result, kindergartners were forming the "intersection of the green blocks and the square blocks." Even without the language of set theory, most children would have understood that what was meant was the square, green blocks.

Set theory in school was a brief episode in Germany. However, one is always searching for ways to improve mathematics education in the schools, for at present, most students leave school with an abiding distaste for mathematics and indeed, not a one of them really understands what it's all about.

Sherlock Holmes Is Confused

In order to understand Russell's paradox, one has to know only what is meant by the expression "x is an element of M" when M is a set. Thus, for example, "14 is an element of the set of even numbers" and "11 is an element of the set of prime numbers" are valid assertions, while "3/14 is an element of the set of integers" is not.

Russell considered the set of those sets that are not elements of themselves. Let us call this set $\mathcal{M}$. It turns out that $\mathcal{M}$ has some strange properties. One could, for example, pose the naive question, is $\mathcal{M}$ an element of $\mathcal{M}$? There are two possibilities for the answer:

- If the answer were "yes," that is, if $\mathcal{M}$ were an element of $\mathcal{M}$, that would imply that $\mathcal{M}$ has the property that characterizes the elements of $\mathcal{M}$: it is not an element of itself, and so the answer "yes" implies the answer "no."
- So let's try the answer "no." This means that $\mathcal{M}$ does not have the characteristic that identifies elements of $\mathcal{M}$ (namely, to be an element of itself). But if $\mathcal{M}$ is not an element of itself, then it must be in $\mathcal{M}$. That is, the answer must be "yes."

It's enough to make your head spin. This form of reasoning exceeds the powers of logic in the realm of set theory. It would be as though in investigating a criminal case in which the perpetrator had to be one of two persons, A or B, Sherlock Holmes were able to draw the following conclusions from the evidence: If one assumes that A is the perpetrator, then the crime must have been committed by B, and if one assumes the hypothesis that B was the perpetrator, then A must be the guilty party. But that is nonsense.

Russell's paradox came as a great shock to the mathematical community. Today, more than one hundred years later, the usual way to avoid such contradictions is to disallow the construction of sets with a self-referential definition and hence sets whose definition you have to know already in order to define them.

Chapter 9

Quit While You're Ahead?

Imagine a game of chance in which you lose your wager with probability $\frac{1}{2}$ and with the same probability get double your wager. (For example, you could toss a coin, losing your bet on "heads" and winning on "tails.") This is certainly a fair game, but is it possible to beat the odds and use this game to make your fortune? In principle, there are several ways of doing so. The first is impossible for us mere mortals: If one could see into the future and know the result of the coin tosses, one could simply choose to play only when one would win. That happens on average every second game, and one could win quite a bit in a single evening.

The second option is more difficult, and also much less profitable. It is well known among gamblers. The idea is simple. Simply bet one euro. If you win, you are one euro richer. Quit and go home. If you lose, then on the second round bet two euros. If this time luck is with you, you can go home with a net profit of one euro (the four euros won minus the three euros wagered). But if you lose on the second round as well, then wager four euros. Again if you win, you are richer by one euro. The strategy is thus to quit if you win and to double your bet if you lose. Eventually, you will win, and will end up with a net profit of one euro.

There are two drawbacks to this method. First of all, it assumes that your financial resources are limitless (in case you have a large number of losses before you finally win) and also that the house will accept bets of unlimited size. Second, you would have a serious problem if in the middle of a losing streak the croupier were to take a holiday.

It is possible to formulate the ideas of not being able to see into the future and of fairness with mathematical precision. Then it can be proved rigorously that in the case of a limited number of rounds or a limit to the amount wagered, there is no winning strategy. Thus all proposed systems for beating the odds are provably worthless: no (honest) gambler becomes rich without luck.

I Almost Always Win

At this point we should append some definitions that one must know in order to formulate precisely the statement of the previous paragraph, the "stopping-time theorem." In its simplest variant it is concerned with *fair games*, that is, games in which winning and losing balance out in each round. Think of tossing a fair coin: heads you win a euro; tails you lose a euro.

We also require a rule that will determine when we break off play. Here are some examples of such rules:

- Quit after the tenth round.
- Quit when you have won a total of 100 euros.
- Quit when you have racked up three losses.

It should be clear that there is an unlimited number of such "stopping rules." (Mathematicians speak of "stopping times," a term that is one of the most important in modern probability theory.)

Once a stopping rule has been chosen, there will be associated with it an average profit: if one plays according to this rule a large number of times, then one should expect that level of profit on average. The stopping-time theorem now states that this average profit is precisely zero, regardless of the complexity of the stopping rule. That is the case at least if one additionally assumes, realistically, that the amount of wager cannot be of unlimited size.

If the amount of expected profit cannot be altered to one's benefit, one can nevertheless change the level of "perceived luck," in that one at least leaves the casino a winner in the majority of cases. Here is a sample strategy for achieving such a goal:

> Choose the doubling-of-bets strategy as described above and play either until you have won one euro profit or until you have reached the house limit on wagers. Then quit for that day and go home.

To analyze this strategy, let us consider an example. Say that the maximum wager is 1,000 euros. If we are having a bad night, we will wager, unsuccessfully, $1, 2, 4, 8, 16, 32, 64, 128, 256, 512$ euros. This represents ten opportunities of winning, each with probability $\frac{1}{2}$. Thus the probability of achieving such a streak of bad luck is $1/2^{10} = 1/1{,}024$, or roughly one in one thousand. Put another way, in almost every case (namely about 999 out of 1,000), we will leave the casino having made a net profit (though not so spectacular as the loss, since the profit is always only one euro). Of course occasionally it will happen that we lose, and the loss will be large, and so the assertion that on average one will achieve a net profit of zero is correct for this strategy as well.

Postscript: In spring 2006, a large portion of the television-viewing public had the opportunity to learn about the mathematical realities associated with the stopping-time theorem. On *Stern TV* with Günther Jauch, there appeared a certain Mr. G., who maintained that he was in possession of a foolproof gambling strategy. Indeed, he left the casino ten times with winnings in his pocket. But of course that proves nothing, since as we have just seen, one can create odds of winning that are close to one hundred percent. The gentleman was unwilling to accept a wager as to whether his system would pass a rigorous test. (The bet was for the author of this book's Christmas bonus.)

Chapter 10

Can a Monkey Create Great Literature?

Let us begin with a thought experiment. Your toddler daughter has sat herself down at the computer and started pounding away on the keyboard. If she keeps at it long enough, every now and then a meaningful word will appear. Would we say that your daughter knows how to write? This question touches on a philosophical problem that received a great deal of attention in the early days of probability theory. Back then there were no personal computers, and the image was one of a monkey at a typewriter (see Figure 1). It can be proved rigorously that if one were to give this monkey sufficient time, it would sooner or later produce every work of literature that has ever been brought to paper. This is because in a sequence of random experiments, everything will eventually occur that has a positive probability: everything that can possibly occur, will eventually (given enough time) occur.

Let us take as an example this very Chapter 10 that you are reading. Even it will eventually appear as a random product of the monkey's keyboarding. The question is whether as a result of this fact one must credit the monkey with a certain degree of creative ability. The answer to this conundrum is not an easy one, since after all, the monkey did actually write the chapter, just as it also will have written Goethe's *Faust* and the lead article in today's newspaper.

Figure 1. A poem? A novel?

There are two reasons to believe that randomness will not replace human creativity. The first argument is a temporal one. What is the use of the certainty that great works of art will sooner or later appear when a rough calculation reveals that enormous eons of time are necessary. Even vast armies of typing monkeys will almost certainly not produce even Part I of *Faust*, even if they stick to their keyboards for millennia. But the second reason is the decisive one. Who is there to declare, "this is it!" when something of lasting value finally appears? Without involvement of a discerning intelligence, there will be no one to separate the meaningful products from the heap of random data. Even you probably don't know whether your daughter has written a fabulous poem in Swahili.

How Much Time Does the Monkey Require?

We would like to estimate how long we might have to wait for something sensible to come out from our monkey's exertions. We begin with a result from probability theory: If a random event has probability p of occurring on a single trial, then one must make on average $1/p$ attempts before the first success. Thus, for example, one expects a card drawn at random from a standard deck of cards to be the king of clubs one time out of 52. Thus the average wait time is the reciprocal of $1/52$, namely 52.

Let us suppose that we are waiting for the word "BIRD." To make our calculations somewhat easier, let us allow our monkey to type four characters; we will then check whether the word "BIRD" has been typed, and if not, put a fresh piece of paper in the typewriter. If the monkey types at random pressing only the letter keys, and we do not distinguish between uppercase and lowercase ("bIRd" counts as a success), then there are twenty-six possible outcomes to each keystroke. Therefore, four keystrokes will produce one of $26 \cdot 26 \cdot 26 \cdot 26$ possible "words." That is a total of 456,976 words, and so the probability that "BIRD" was typed is $p = 1/456{,}976$, and the expected number of attempts is 456,976. That is the order of magnitude that one should plan on, though of course, that is an average value. The actual number of attempts in any particular case could be much less or much more.

So what does it all mean? If we give the monkey a new sheet of paper every ten seconds, then it will be able to make 6 attempts per minute, which makes 360 per hour, and so $8 \cdot 360 = 2{,}880$ per eight-hour workday. Now we must divide 456,976 by 2,880 to determine the expected number of days that it will take before "BIRD" appears. The result is approximately 159, and so for "BIRD" to appear we would likely have to wait about half a year.

Now, "bird" is not a very complicated work of literature. What if instead we wish to wait for the phrase "SEVEN BIRDS IN THE HAND"? That is a total of 23 characters. Now we have to take into account the space character, so there are now twenty-seven different possibilities for each keystroke, so each outcome has probability 1/27. Therefore, the probability that twenty-three random keystrokes will produce "SEVEN BIRDS IN THE HAND" is

$$\frac{1}{27^{23}} = \frac{1}{834{,}385{,}168{,}331{,}080{,}533{,}771{,}857{,}328{,}695{,}283}.$$

The expected number of attempts is therefore

$$834{,}385{,}168{,}331{,}080{,}533{,}771{,}857{,}328{,}695{,}283,$$

which represents a "working time" of about 10^{28} years if again we begin a new attempt every ten seconds and specify an eight-hour workday. It is not likely that our monkey will achieve success.

Chapter 11

The Birthday Paradox

It has been mentioned in this book that human intuition is not particularly well adapted to grasping mathematical truths; evolution required the internalization of only very elementary facts relating to "space" and "number." This is especially true with respect to the branch of mathematics known as probability theory, where expectation and mathematical fact are frequently at odds.

A famous example of this phenomenon goes under the name "birthday paradox." Suppose that twenty-five people have gathered for a party. Is it likely that two of these individuals have the same birthday? What are the odds? This probability can be calculated without too much difficulty, and in fact, it is surprisingly about 57%.

If one repeats the question for various numbers n of partygoers, one sees that the probability of two people sharing a birthday becomes quite large for values of n that are not very large at all. The number 23 plays a special role here. It is the smallest number for which the probability of two people sharing a birthday is greater than 50%. This goes against intuition; most people guess that the 50% mark is reached at 183 people, which is approximately half of 365.

Those who don't particularly trust mathematics can convince themselves by a bit of research. If you have a child of elementary-school age, you have only to look at the birthday calendar of a couple of classes at the next parents' night. It is rather the rule than the exception that you will find at least one date on which two pupils share a birthday.

From a formal point of view, one can solve the birthday paradox by calculating the probability that among n randomly selected numbers between 1 and 365, at least two are the same. If one replaces 365 by a different number, the calculation is equally as simple, and there are new and interesting interpretations. For example, the probability that in a randomly generated seven-digit telephone number at least one number appears twice is 94%. (In this case, seven random numbers are chosen from the set $0, 1, \ldots, 9$.) Might that not be a good starting point for making a small wager? For example, I might bet—without much risk—that at least one number appears twice in your telephone number.

How Are These Probabilities Determined?

The general formulation of the problem is this: Given n objects, choose an object at random from the entire collection r times. Each object has the same probability of being chosen, and each can be conceivably selected more than once.

Example 1, birthdays: Here the "objects" are possible birthdays, that is, $n = 365$. Then r is the number of guests at the party, and the distribution of birthdays is interpreted as "select from the possible birthdays."

Example 2, words: If one types an r-letter word at random, then this can be formulated as the case $n = 26$ of the selection problem.

Example 3, telephone numbers: This corresponds to the case $n = 10$ (since there are ten digits) and $r = 7$ (for seven-digit numbers).

The problem is now to calculate the probability that all the chosen objects are different from one another. If one knows that, then one knows as well the probability that at least two are alike: one has only to compute "one minus the probability that they are all different." For example, if the probability that all the birthdays are unique is 0.65, then the probability that "at least two birthdays are the same" is $1 - 0.65 = 0.35$, that is, 35%.

To solve the problem, we invoke the following principle:

probability = favorable cases divided by possible cases.

This principle comes into play whenever all the possible objects are selected with the same probability. The number of possible cases, that is, the number of possible selections, is equal to n^r, the product n times n times n ... a total of r times. (The reason is that for each of the r selections, there are n possibilities.)

Now to the "favorable" cases: How many selections are there in which all the objects are different? For the first selection not much can happen, since there are n possibilities. For the second, one must avoid choosing the first-selected object, and so there are only $n-1$ possibilities for the second, and thus $n \cdot (n-1)$ for the first two. For the third selection, there are now two forbidden elements, and we arrive at $n(n-1)(n-2)$ possibilities of selecting three objects without repetition.

And so it goes; for r selections the number is

$$n(n-1)(n-2)\cdots(n-r+1).$$

Therefore, to compute the quotient "favorable cases divided by possible cases," we must evaluate the quotient

$$\frac{n(n-1)(n-2)\cdots(n-r+1)}{n^r},$$

which can be reconfigured as[1]

$$1 \cdot \left(1-\frac{1}{n}\right) \cdot \left(1-\frac{2}{n}\right) \cdots \left(1-\frac{r-1}{n}\right).$$

It should now be clear how the results given above were obtained. The number 23 comes from the fact that the probabilities for "no sharing of birthdays among r individuals" first slips below the 0.5 barrier at $r = 23$. Indeed,

$$\left(1-\frac{1}{365}\right)\left(1-\frac{2}{365}\right)\cdots\left(1-\frac{22}{365}\right) = 0.493$$

[1]The trick is to rearrange the factors: write the fraction as

$$\frac{n}{n} \cdot \frac{n-1}{n} \cdots \frac{n-r+1}{n},$$

and then simplify each factor using the relation $\frac{x-y}{x} = 1 - \frac{y}{x}$.

is less than 0.5, while

$$\left(1-\frac{1}{365}\right)\left(1-\frac{2}{365}\right)\cdots\left(1-\frac{21}{365}\right)$$

is equal to 0.524.

Since $1 - 0.493 = 0.507$, the probability of at least two among twenty-three partygoers sharing a birthday is 50.7%. And see what happens as the party gets larger: with thirty guests the probability is 71%; with forty, it is 89%; and with 50, it is 97%.

These probabilities grow faster than one would have naively expected. To illustrate the phenomenon, we present two tables.

The first table involves *agreement of digits*: What is the probability that among r randomly selected digits, at least two are the same? The number r appears in the first row. In the second row is the probability that all r randomly selected digits are different, while row three gives the probability that at least two are the same:

1	2	3	4	5	6	7	8	9	10
1.000	0.900	0.720	0.504	0.302	0.151	0.060	0.018	0.004	0.0004
0.000	0.100	0.280	0.496	0.698	0.849	0.940	0.982	0.9964	0.9996

If you would like to know, say, the likelihood that in a seven-digit telephone number two digits are the same, just look at the column for $r = 7$: the probability is surprisingly high, namely 0.94.

Next we offer the table belonging to the birthday paradox. The first row contains the number of partygoers; the second, the probability that all of them have different birthdays; and the third, the complementary probability (that is, the probability that at least two of them have the same birthday):

1	2	3	4	5	6	7	8
1.000	0.997	0.992	0.984	0.973	0.960	0.944	0.926
0.000	0.003	0.008	0.016	0.027	0.040	0.056	0.074

9	10	11	12	13	14	15	16
0.905	0.883	0.859	0.833	0.806	0.777	0.747	0.716
0.095	0.117	0.141	0.167	0.194	0.223	0.253	0.284

17	18	19	20	21	22	23	24
0.685	0.653	0.621	0.589	0.556	0.524	0.493	0.462
0.315	0.347	0.379	0.411	0.444	0.476	0.507	0.538

Are All the Dice Different?

We would like to mention a special case of the birthday paradox: if n selections are made from an n-element set, then the probability that each of the n elements is selected exactly once is $n!/n^n$. (Recall that $n!$ is shorthand for $1 \cdot 2 \cdots n$.) This is the number that arises if in the previous examples we were to take $r = n$.

Example 1: If an integer from the set $1, 2, \ldots, 9$ is selected nine times, then the probability that all nine integers are different is

$$\frac{9!}{9^9} = \frac{362{,}880}{387{,}420{,}489} = 0.000936\ldots,$$

that is, just under one-thousandth.

Example 2: If six dice are thrown simultaneously, the probability that all of them come up with a different number is given by

$$\frac{6!}{6^6} = \frac{720}{46{,}656} = 0.0154\ldots.$$

This is about the same as the probability for three correct in the lottery (see Chapter 40). Thus on average, one should expect six different numbers to appear one time out of sixty-five.[2]

Afterword: The German national soccer team had twenty-three players on its squad in 2006. There were thus good odds for a double birthday. And indeed, Mike Hanke and Christoph Metzelder celebrate their birthdays on 5 November.

[2]If the probability of success of a random event is p, then $1/p$ is the expected number of attempts before the first successful event, and $1/0.0154\ldots \approx 65$.

Chapter 12

Horror Vacui

Students of mathematics have a great respect for nothingness, at least at the beginning of their studies. That is not surprising, since it took centuries for the number zero to be accepted on equal terms with numbers like seven and twelve. To understand the difficulty, one must recall that set theory, as developed by Georg Cantor, is the foundation of modern mathematics. According to Cantor's theory, a set is a collection of certain distinct objects joined into a new object. This concept is nothing strange to nonmathematicians, who understand that, for example, the USA is a set comprising fifty different states, or that the European Union is a set of twenty-seven member states.

Things become problematic when in forming such a collection, there is nothing to collect, for example, the set of all German citizens who are more than ten feet tall. It is not easy to comprehend that the object thus defined is the same as that defined by the condition "all Ukrainian pianists who can play Chopin's *Valse Minute* in twenty seconds." Both are examples of the *empty set*.

The empty set (denoted by $\varnothing$) plays a role in set theory analogous to that of zero for the integers. One can add the empty set to an arbitrary set without changing its size, and it is this property that characterizes this set. All of mathematics can be developed from the empty set. For example, the integers arise by taking the empty set as zero and then the number 1 as the set whose single element is the empty set. This becomes somewhat messy for larger numbers, but the principle is the same.

Having an understanding of what is meant by the empty set by no means solves all of our difficulties. Mathematicians study properties of sets, and of particular importance are assertions of the form, "all elements of the set have such-and-such property." It is usually taken as given that all such statements are true for the empty set.

> As in most cases, here one has actually only to apply everyday logic. For example, if someone has made the promise to give five euros to every beggar he sees that day, he has fulfilled his obligation even if he does not see a single beggar.

The Empty Set Behaves like Zero

Here we would like to clarify what is meant by the statement that the empty set corresponds to the number zero. First, one must know that the *union* of two sets A, B consists of all the elements that are in A or in B (or in both sets). See Figure 1. For example, if A consists of the elements $2, 5, 6$ and B consists of $6, 8$, then the union of A and B is the set comprising $2, 5, 6, 8$. Or if A is the set of residents of Paris and B is the set of blond Frenchmen, then the union of the two sets includes, among others, Parisian brunets and Parisian redheads.

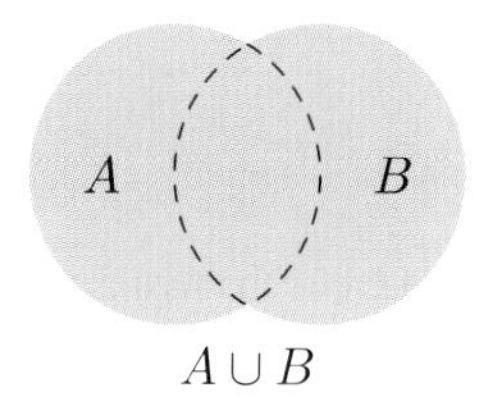

Figure 1. Thus can one picture the union of two sets.

The notation for the union of two sets is $A \cup B$ (read "A union B"). Then for every set A we have the relation

$$A \cup \varnothing = A,$$

since the empty set offers no new elements to add. If we think of set union as the analogue of addition, then the relation $A \cup \varnothing = A$ corresponds to the equation $x + 0 = x$ for numbers.

We also have the notation $A \cap B$ for the set of elements that are contained in both A and B (in the example above, the result is the set of blond Parisian men). See Figure 2.

The expression $A \cap B$ is read "A intersect B."

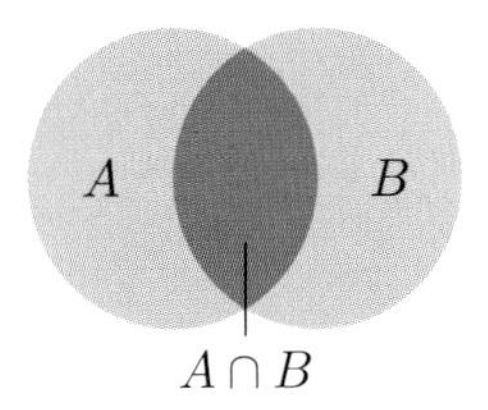

Figure 2. And this is what the intersection looks like.

Since the empty set has no elements, we have

$$A \cap \varnothing = \varnothing.$$

As we thought of set union as the analogue of addition, set intersection behaves similarly to multiplication. For example the above relation for sets corresponds to the equation $x \cdot 0 = 0$ for numbers.

"Horror vacui" is Latin for "abhorrence of emptiness." This notion comes from ancient natural philosophy, which held that "nature abhors a vacuum," that is, that empty space cannot exist. And indeed, while vacuums can and do exist, they do have a tendency to "suck in" gases or liquids.

Chapter 13

Sufficient Difficulties with the Logic of Mathematics Are in Fact a Necessity

Today's topic is the human species' logical capabilities. To bring some order into the many impressions on the subject that we experience daily, let us attempt to construct some logical relationships. Consider, for example, the obviously true sentence, "If today is Christmas, there will be no mail delivery." Nobody confuses this with the converse statement, "If there is no mail delivery today, it must be Christmas." Remarkably, the temptation is great to confuse such pairs of assertions. One need only think of the "clothes make the man" phenomenon. Wealthy folks can dress themselves up in fine clothes, but one cannot easily deduce from a woman's dress the state of her bank account.

Things become more difficult to explain when the situation becomes more abstract. In this regard, the "trapezoid controversy" comes to mind. In February 2003, the entire nation was wrapped up with the problem—in connection with the "who will be a millionaire question" of whether a rectangle is a trapezoid. If one accepts the definition "a trapezoid is a quadrilateral in which two sides are parallel," then the answer should be yes. This was difficult to explain to a host of respectable citizens, whose reactions ranged from the malicious to the aggressive: How can a mathematician assert that every trapezoid is a rectangle? But no one said any such thing....

For the sake of completeness, let us add this: If for two assertions p and q it is always the case that "q follows from p," then mathematicians say that "p suffices for q," and that "q is necessary for p." It is easy to confuse these two concepts. Or can you determine whether the following statement is correct: "for a figure to be a trapezoid, it suffices that it be a rectangle"? (The correct answer is contained in the footnote below.)[1]

Trapezoid or Not?

In an attempt to avoid divorce and fisticuffs among the trapezoid disputants, one must admit that there is quite a bit of confusion in textbooks and lexicons. Indeed, one often finds in the definition of a trapezoid the condition that there be no right angle.

From a mathematical point of view such a restriction makes little sense, since it is very uneconomical. Consider, for example, the fact that the sum of the angles of a trapezoid is 360°. Suppose that we have convinced ourselves of this via a rigorous mathematical proof. Now we study the concept of "rectangle." If we, like all mathematicians, consider rectangles to be a special case of trapezoids, then we can immediately enunciate the theorem that "the sum of the angles of a rectangle is 360°, since it is a special case of the more-general result that we have already proved. Everyone else must start over from scratch. And the same holds for every future result about trapezoids.

Most school textbooks picture a trapezoid like the one in Figure 1. But the drawings in Figure 2 are also trapezoids.

[1] Indeed, "the figure is a rectangle" is *sufficient* for the correctness of the assertion, "the figure is a trapezoid."

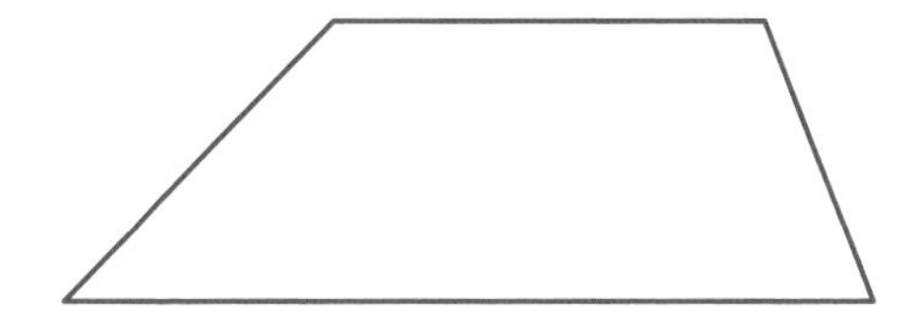

Figure 1. A "typical" trapezoid.

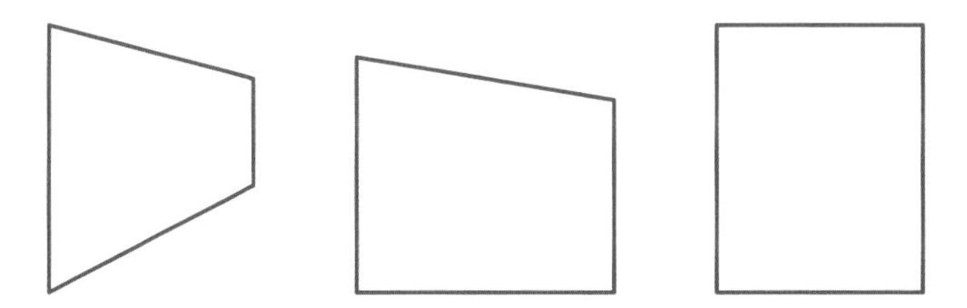

Figure 2. Additional examples of trapezoids.

Do Dogs Think Logically?

Figure 3 shows a sign on a door that reads, "Dogs that bark do not bite. Our dog does not bark." It is surely meant (humorously?) as a warning. What has happened is that from "p implies q" ("barks" implies "does not bite") the conclusion was drawn that "not p implies not q" ("doesn't bark" implies "bites"). This is not a logical conclusion, but the sign nevertheless fulfills its purpose.

Figure 3. Would you walk through this door?

Chapter 14

To Change or Not to Change? The Monty Hall Problem

Probability theory is rich in paradoxes; there are many valid statements that contradict "normal human reasoning." It is thus that a few years ago, the so-called Monty Hall problem became known to the public at large.

In case you have forgotten, here is what it was all about: The quizmaster of a popular game show invites the finalist on the show to choose one of three doors. Behind one of them lies the grand prize: a new car. Behind each of the other two is a booby prize, in the form of a goat. Say the contestant chooses door number 1. The quizmaster then opens, say, door 3, behind which stands a goat. We now get to the point: The contestant is given the opportunity to change doors, in this case from door 1 to door 2. Should he accept the offer? In favor of nonacceptance is the fact that the location of the grand prize has not changed by the opening of door 3, while an argument for acceptance is that by opening the door, the quizmaster has altered the situation. The question of the correct answer caused disputes among the various branches of mathematics; the problem found its way into the major newspapers and was discussed intently by nonmathematicians. The "yes" faction took the ideas of the "no" faction for naive, laughable, primitive, and of course vice versa. The matter also had a gender-specific aspect.

One of the first advocates for accepting the offer to change doors was the American journalist Marilyn Vos Savant, a woman who became famous for an unusually high IQ. There were more than a few voices from the mathematical community advising her to steer clear of the matter, not to get involved in affairs that she, as a lay woman, could not possibly understand.

Who was correct? Indeed, Marilyn was right that the contestant should accept the switch, since his chances of winning increase from $\frac{1}{3}$ to $\frac{2}{3}$. The reasoning for this assertion follows below.

Analysis: Why Is It More Favorable to Switch?

The assertion that choosing the other door that is offered is more favorable is only a first approximation to the truth. We offer a detailed analysis of the problem so that this assertion can be understood and the entire situation made clear. The path to this goal is not entirely simple, because it involves a rather complex phenomenon.

Probability. First, we must concern ourselves with a few concepts from probability theory. Fortunately, we do not need to resolve the question "what is probability?"

Let us imagine a procedure that produces random events. For example, one could toss a die or draw a card from a well-shuffled deck. If one repeats this process a large number of times, one observes certain "tendencies." In about one-sixth of the throws of the die, a four appears, while in about one-fourth of the cases, a heart is drawn from the deck of cards. One expresses this phenomenon by saying that in the case of the die, the probability of throwing a four is one in six, or one-sixth, and for the card-drawing, that the probability of drawing a heart is one in four, or one-fourth. In general:

> The probability of a possible result E in a random drawing is the number p having the following properties: if one repeats the random selection very often, then in a fraction p of the cases, the result E will be achieved. While this is only approximately true, it becomes more and more accurate, the more repetitions of the selection occur. One then writes $P(E) = p$, which is read, "the probability of E is equal to p."

In our examples, we have $P(\text{throw a four}) = \frac{1}{6}$ and $P(\text{draw a heart}) = \frac{1}{4}$. Since a part of the whole is always between zero and one, the same is true of probabilities. Furthermore, the definition of probability makes some other simple properties clear. For example, if a result E always fulfills the condition that one has set for a result F, then the probability of F is at least as great as that for E. There is no way that it can be smaller. For example, since 4 is an even number, it is no surprise that the probability of throwing an even number (which is 0.5) is greater than that of throwing a four.

In the Monty Hall problem, a number of probabilities enter the picture. For example, it would be of interest to know the probabilities with which the grand prize of a new car is placed behind the various doors. Can one assume the same probability for each of the three doors (that is, $\frac{1}{3}$)? Or does the door that is closest to the stage entrance have a higher probability? (After all, it is not so easy to push a large automobile onto the stage.)

Conditional Probabilities. We now discuss the important principle, "information changes probabilities." An example: The probability of rolling a four is one-sixth. But if after the die is cast but before you look at the number, you are informed that the number that came up is even, the situation is quite different. It is now clear that the die can show only one of the three numbers two, four, six, and therefore the probability of it being a four rises to one-third. Or perhaps the additional information is that the number is odd. Then it is certainly not a four, and the probability has shrunk to zero.

In sum, any probability can arise if one has obtained additional information. It can remain the same, increase, or decrease.

We experience the same phenomenon in our everyday lives.[1] Suppose that every workday you commute along the same road. The traffic is moving somewhat faster in the left lane, and you would therefore like to shift into that lane. It would thus be of interest to know whether the car in front of you is about to turn left at the next intersection. (If so, you should change lanes, since you want to continue straight ahead.) Suppose that about one in every twenty drivers turns left at that intersection, and therefore you can estimate the probability of a left turn as $1/20$. But all German license numbers include information about the city in which the car is registered, and it could happen that the license plate of the car in front has a number indicating that it is from a town that is reached by turning left at this intersection. In such a case, the probability that the car will make a left turn has certainly increased.

It will be useful to formulate all of this more precisely. If E is an event, then $P(E)$ denotes the probability that E occurs. And if F is some additional information, then one writes $P(E \mid F)$ for the new probability that E occurs given the information F. One reads this as "the probability of E given F," and the number $P(E \mid F)$ is called the "conditional probability of E given F."

In the introductory example, E was the result that a four was rolled, and F the information that the number rolled was even. We convinced ourselves that in this example, $P(E \mid F) = \frac{1}{3}$.

In general, one proceeds as follows. First, one determines $P(F)$ (the probability of F) and $P(E \text{ and } F)$ (the probability that both E

[1] I would go as far as to suggest that the adaptation of probability to new information has played an important role in the evolutionary development of the human species, eventually becoming "hard-wired" in our brains.

and F occur). Then $P(E \mid F)$ is defined as follows:

$$P(E \mid F) := \frac{P(E \text{ and } F)}{P(F)}.$$

Let us test this on our example. We have $P(F) = \frac{1}{2}$, since one-half of the rolls of a die produce even numbers. Next, "E and F" corresponds to the result that a number occurs that is both even and equals four. That happens only if a four is rolled, and therefore $P(E \text{ and } F) = \frac{1}{6}$. We therefore obtain

$$P(E \mid F) = \frac{P(E \text{ and } F)}{P(F)} = \frac{1/6}{1/2} = \frac{1}{3}.$$

As a further example, consider choosing a card at random from a standard 52-card deck. The probability of E = the queen of diamonds is $1/52$, since there is only one such card in the pack. However, if someone were to have a peek at the card and tell you that it was a queen, then the probability that it is the queen of diamonds rises to $1/4$: With F = queen, we have $P(F) = 4/52 = 1/13$ (there are four queens) and $P(E \text{ and } F) = 1/52$ (there is only one queen of diamonds). It then follows that

$$P(E \mid F) = \frac{P(E \text{ and } F)}{P(F)} = \frac{1/52}{1/13} = \frac{1}{4}.$$

The Bayesian Formula. Remarkably, conditional probabilities can be more-or-less inverted. To do so, we use something called the Bayesian formula. Here is a real-life example:

> Shortly after a visit from some friends, you notice that your favorite DVD has disappeared. You know that one of your friends has a tendency every now and then to "borrow" items without asking. Whom should you suspect?

For our purposes it is best to formulate the Bayesian formula as follows: Consider a random experiment in which every outcome falls into one of three classes, which we name B_1, B_2, B_3. It is crucial that these classes have no overlap.

Taking an example from dice-rolling, we could define the classes as follows:

> B_1: A 1 or a 2 was rolled.
>
> B_2: A 3 or a 4 was rolled.
>
> B_3: A 5 or a 6 was rolled.

We now consider a possible result of our experiment, which we will call A. For example, we could have "$A =$ a prime number was rolled." If the conditional probabilities $P(A \mid B_1)$, $P(A \mid B_2)$, $P(A \mid B_3)$ are known as well as the probabilities $P(B_1)$, $P(B_2)$, $P(B_3)$, then we can recover the "inverse" conditional probabilities, that is, $P(B_1 \mid A)$, $P(B_2 \mid A)$, $P(B_3 \mid A)$ as follows:

$$P(B_1 \mid A) = \frac{P(A \mid B_1)P(B_1)}{P(A \mid B_1)P(B_1) + P(A \mid B_2)P(B_2) + P(A \mid B_3)P(B_3)},$$
$$P(B_2 \mid A) = \frac{P(A \mid B_2)P(B_2)}{P(A \mid B_1)P(B_1) + P(A \mid B_2)P(B_2) + P(A \mid B_3)P(B_3)},$$
$$P(B_3 \mid A) = \frac{P(A \mid B_3)P(B_3)}{P(A \mid B_1)P(B_1) + P(A \mid B_2)P(B_2) + P(A \mid B_3)P(B_3)}.$$

These are called *Bayesian formulas.*[2]

We have no intention of proving this formula here. However, we shall provide an example as illustration. The classes B_1, B_2, B_3 will be as above, namely $B_1 =$ "1 or a 2 was rolled," $B_2 =$ "3 or a 4 was rolled," $B_3 =$ "5 or a 6 was rolled." For the event A we shall choose "the number rolled is greater than 3." Carrying out the calculations described in the previous section, we obtain

$$P(A \mid B_1) = 0,$$
$$P(A \mid B_2) = \frac{1}{2},$$
$$P(A \mid B_3) = 1,$$

and we clearly have $P(B_1) = P(B_2) = P(B_3) = \frac{1}{3}$.

[2] If instead of three classes one has the general number n, that is, $B_1, B_2, \ldots, B_n$, then the Bayesian formulas take the form

$$P(B_i \mid A) = \frac{P(A \mid B_i)P(B_i)}{P(A \mid B_1)P(B_1) + \cdots + P(A \mid B_n)P(B_n)},$$

where one may replace the symbol i by any one of $1, 2, \ldots, n$.

The die is cast, and the result in fact is in A (the number rolled is greater than 3). What is the probability that it is in, for example, B_2? To answer this question, we use the Bayesian formula:

$$\begin{aligned} P(B_2 \mid A) &= \frac{P(A \mid B_2)P(B_2)}{P(A \mid B_1)P(B_1) + P(A \mid B_2)P(B_2) + P(A \mid B_3)P(B_3)} \\ &= \frac{(1/2)\cdot(1/3)}{0\cdot(1/3) + (1/2)\cdot(1/3) + 1\cdot(1/3)} \\ &= \frac{1}{3}. \end{aligned}$$

Analogously, using the Bayesian formula, one could calculate[3] that $P(B_1 \mid A) = 0$ and $P(B_3 \mid A) = \frac{2}{3}$.

The Best Strategy for the Monty Hall Problem: Standard Variant. After these extensive preparations, we can now determine whether changing doors is advantageous. We begin with the probabilities concerning the door behind which the car is hidden. We will use the following shorthand:

B_1**:** The car is hidden behind door 1.

B_2**:** The car is hidden behind door 2.

B_3**:** The car is hidden behind door 3.

We are going to assume that each of these possibilities is equally likely, and thus that

$$P(B_1) = P(B_2) = P(B_3) = \frac{1}{3}.$$

Perhaps that is a bit naive, but if there is no information to the contrary, we may begin with such an assumption.

Now comes the big moment at which the decision must be made. The contestant has chosen door 1, the quizmaster reveals the goat behind door 3, and it is not clear whether one should switch one's allegiance from door 1 to door 2. Here is the analysis.

[3] In this case a direct calculation would be easier, yielding of course the same result.

The event "the quizmaster reveals a goat behind door 3" will be abbreviated A. Using this information, we would like to answer the question, is the auto more likely behind door 1 or door 2? Using the notation that we have developed, we would like to know the relationship between the numbers $P(B_1 \mid A)$ and $P(B_2 \mid A)$. If they are the same, then switching offers no advantage, but if the second is larger, then one should switch.

This is a typical application of the Bayesian formula. To apply it, we require the numbers $P(A \mid B_1)$, $P(A \mid B_2)$, $P(A \mid B_3)$.

How large is $P(A \mid B_1)$? In words, if the car is behind door 1, what is the probability that the quizmaster opens door 3? He could of course also open door 2 (or simply tell which door conceals the car). We are going to assume here that he opens door 2 or door 3 with equal probability, and therefore we set $P(A \mid B_1) = \frac{1}{2}$.

It is easier to determine $P(A \mid B_2)$. If the car is behind door 2, what is the probability that door 3 will be opened? It is clearly 1, since the quizmaster cannot open door 2 (the car is there), and door 1 is also taboo, since that is the door chosen by the contestant.

It is similarly easy to deal with $P(A \mid B_3)$. It is certainly zero, since he will certainly not open the door that conceals the car. Collecting this information, we have

$$\begin{aligned} P(A \mid B_1) &= \frac{1}{2}, \\ P(A \mid B_2) &= 1, \\ P(A \mid B_3) &= 0. \end{aligned}$$

And now the Bayesian formula can be applied:

$$\begin{aligned} P(B_1 \mid A) &= \frac{P(A \mid B_1)P(B_1)}{P(A \mid B_1)P(B_1) + P(A \mid B_2)P(B_2) + P(A \mid B_3)P(B_3)} \\ &= \frac{(1/2) \cdot (1/3)}{(1/2) \cdot (1/3) + 1 \cdot (1/3) + 0 \cdot (1/3)} \\ &= \frac{1}{3}, \end{aligned}$$

while

$$P(B_2 \mid A) = \frac{P(A \mid B_2)P(B_2)}{P(A \mid B_1)P(B_1) + P(A \mid B_2)P(B_2) + P(A \mid B_3)P(B_3)}$$
$$= \frac{1 \cdot (1/3)}{(1/2) \cdot (1/3) + 1 \cdot (1/3) + 0 \cdot (1/3)}$$
$$= \frac{2}{3}.$$

Since $P(B_1 \mid A)$ is the probability of winning the car with the strategy "don't switch" and $P(B_2 \mid A)$ is the probability of winning the car with the strategy "switch," it is now clear that switching doubles the chances of winning the car.

The Monty Hall Problem: The Whole Truth. If you have been following the foregoing analysis attentively, you will have taken note

that to achieve the outcome "switching is better; it doubles your chances of winning," certain assumptions were necessary. For example, we set $P(A \mid B_1) = \frac{1}{2}$. That is not the only possibility. Perhaps the quizmaster always opens door 3 whenever possible (that is, when it does not conceal the car). To investigate the general case, we set $P(A \mid B_1) = p$, where p is some number between 0 and 1. Then our previous analysis yields

$$P(B_1 \mid A) = \frac{p}{1+p}, \quad P(B_2 \mid A) = \frac{1}{1+p}.$$

Since $p < 1$, the first of these two numbers is always smaller than the second, and so we see that it is always more favorable to switch, though the improvement in the probability could be quite small.

There is another way of tackling this problem.[4] The contestant employs a strategy of paying no attention to the intention of the quizmaster and simply always switches. The argument is as follows:

- At the time of the initial choice (and this holds whether or not one eventually switches doors), one has won the car with probability $\frac{1}{3}$, since the car was placed behind each door with equal probability.

[4] My thanks to Professor Dieter Puppe, of Heidelberg, for the idea.

- In switching, one wins precisely when the initial choice was incorrect, and therefore with probability $\frac{2}{3}$.

This argument can be refined a bit. Let p_1, p_2, p_3 denote the probabilities that the auto is hidden behind door 1, door 2, door 3. Then if door 1 is chosen, the probability of "the car is won without switching doors" is equal to p_1, and the probability of "the car is won with switching doors" is equal to $p_2 + p_3$.

It is possible that some readers might be puzzled by the fact that in the second analysis the actions of the quizmaster apparently play no role. One must consider the matter with care to see that both analyses are correct.

In the first analysis the initial situation was given thus: Door 1 is chosen, door 3 (with the goat) is opened. And from this one has to determine the relevant probabilities.

In the second analysis the situation was different. The action of the quizmaster was irrelevant, and one should switch in any case. Nevertheless, it is intuitively difficult to see that this different information is responsible for the different probabilities.

Chapter 15

In Hilbert's Hotel There Is Always a Vacancy

Mathematicians often have to deal with infinity. Experience has shown that in the realm of the infinite things happen that are quite different from what occurs in our finite experience.

For what follows, it will suffice to consider the simplest infinite set, namely the natural numbers $1, 2, 3, \ldots$. The great Galileo, in his 1638 *Discorsi*, marveled over the strange phenomena that can occur in this realm. There it is determined that there are just as many natural numbers as there are square numbers, that is, the numbers $1, 4, 9, 16, \ldots$, since one can write down both sequences one above the other and thereby obtain a one-to-one correspondence between them. The mathematical underpinning is this: When one removes part of an infinite set, what remains can have the same "size" as the original set.

The mathematician David Hilbert (1862–1943) came up with an interesting way of describing this phenomenon, which became known as "Hilbert's Hotel." See Figure 1. This hotel has infinitely many rooms, numbered $1, 2, 3, \ldots$. It is a vacation weekend, and the hotel is full. Late in the evening a visitor arrives and asks for a room. Normally, a full hotel means no vacancy, but not in Hilbert's Hotel, where there is a solution to the problem: the guest in room 1 moves

Figure 1. Hilbert's Hotel.

to room 2, the guest in room 2 moves to room 3, etc. And now room 1 is free, and everyone can have a restful night. But later that night, a minivan containing eight vacationers seeking rooms arrives. Still the problem can be solved: the guest in room 1 can move to room 9, and so on.

The systematic study of infinite sets began only in the nineteenth century with the pioneering work of Georg Cantor (1845–1918). For his work he was given the cold shoulder by many of his mathematical colleagues, who believed that mathematics should be restricted to the concrete and constructible. Today, Cantor has been fully rehabilitated, and the infinite belongs to a mathematician's everyday toolbox as much as the integers, geometric objects, and probabilities.

It Was a Hectic Night!

The events of that turbulent night at Hilbert's Hotel continued; there is much more to tell. At a nearby railroad station a train arrives containing *infinitely many* passengers, each of whom would like a hotel room. They are tired, and moreover, they all have reservations.

But the hotel is full. What is to be done? No problem. The automated notification system informs the guest in room 1 to move to room 2, the guest in room 2 to move to room 4, the guest in room 4 to move to room 6, and so on, each guest moving to the room whose number is double that of the current room. Now all the rooms with odd numbers are free, and all the worn-out train travelers can be accommodated.

Although what we have presented is merely a thought experiment for illustrating some phenomena related to infinite sets, we should mention a practical objection to the above solution. Guest n, currently in room n, is to move to room $2n$. For small values of n, this is not difficult, but for large n, there will be a long distance to travel between the old and new rooms. If we assume that Hilbert's guests are limited in how fast they can walk, it follows that the room-changing cannot take place in a finite amount of time.

But this problem already existed in the initial situation. If all the guests are notified simultaneously of their impending move, then it would work: all could move at the same moment, and after ten minutes, calm would reign again. But since no information can travel faster than light, the guests in far-off rooms will be notified only after a long period of time.

Chapter 16

That Fascinating Number Pi

The number π (say "pie") has an excellent chance of coming in first in any discussion among mathematicians about the most important number of all. Its significance for geometry is well known, and formulas such as "the circumference of a circle is π times the diameter" are inculcated in schoolchildren.

But this number shows up in virtually every area of mathematics, even where there does not appear to be a circle in sight. That it is important for probability theory could be seen by anyone taking a glance at a ten-mark note (see Chapter 25), where it was chosen as an illustration of the contributions of the great mathematician Gauss.

As a number, π has many fascinating properties. If one wishes to employ it in a concrete formula, say for the number of seeds necessary for planting a circular field, one may use an approximation to a small number of decimal places, for example, $\pi \approx 3.14$. However, it can be proved that no number of digits after the decimal point can provide an exact value for π. An infinite number are required. Indeed, π is a *transcendental number*, and these are the most complicated in the hierarchy of numbers. This fact was proved in the nineteenth century, and as a corollary, the ancient problem of "squaring the circle" was solved (see Chapter 33 for more on this).

If one cannot write down all of the digits after the decimal point, why not write down as many as possible? There is a competition among some mathematicians and computer scientists to do just that, using computers and refined theoretical results to achieve ever new records. Currently, several billion digits have been recorded. This is not only idle play. The number π hides many secrets, and it is hoped that by analyzing its digits, some of them will be revealed.

We should finally mention that π has exercised a certain fascination over nonmathematicians as well. There are π fan clubs, and a few years ago, a π film appeared. Just the thing for getting in the mood for Givenchy's "π" perfume.

π in the Bible

Anyone wishing to read between the lines can find that the number π is referred to in the Bible (1 Kings 7.23):

> And he made a molten sea, ten cubits from the one brim to the other: it was round all about, and his height was five cubits: and a line of thirty cubits did compass it round about.

Here the "molten sea" is some sort of urn for holy water that was set before the temple of Solomon. If one imagines it as circular in form, then one may take the following information from the biblical text:

> circumference divided by diameter equals 3.

That is a notably bad approximation to π. The Babylonians and Egyptians worked with the much better approximation $\pi \approx 22/7 = 3.142\ldots$. To be sure, the imprecision can be explained, say, in that the urn was measured not at the brim, but somewhat lower down.

π: Some Estimates

Some facts concerning π can be explained without any mathematical concepts. Suppose that we have a circle inscribed in a square, as in the left-hand picture in Figure 1.

If we were to move along the circle beginning at a point where the circle touches the square and ending at the opposite point, we

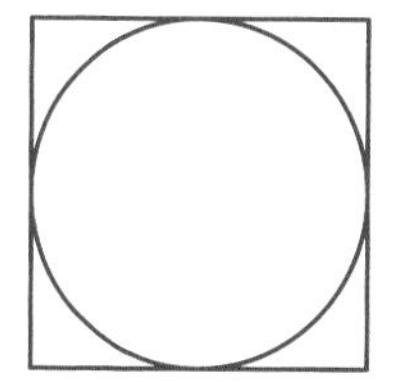 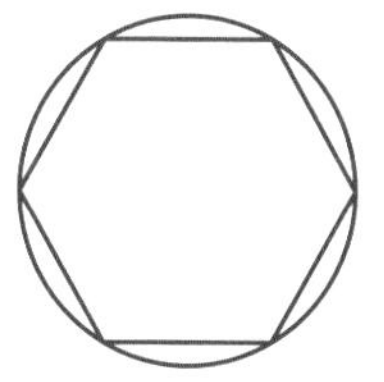

Figure 1. π is smaller than 4 and larger than 3.

would have covered a distance equal to half the circumference of the circle. This distance is therefore equal to one-half the product of the diameter and π, namely $\frac{1}{2} \cdot 2r \cdot \pi = \pi \cdot r$, where r is the radius (and therefore $2r$ is the diameter).

It is also clear that if we traveled around the square between the same two points, then we would cover the distance $4 \cdot r$. We see, then, that $\pi \cdot r$ is less than $4 \cdot r$, and if we divide this inequality by r, we obtain the result that π is less than 4. In a similar way, one can convince oneself that π is larger than 3. This time, we draw a circle around a hexagon (see the right-hand picture in Figure 1). Now the path between the two opposite points is *shorter* if we travel along the edges of the hexagon. We have three edges to traverse, each of length r (the radius of the circle), and that leads us to the fact that $3 \cdot r$ is less than $\pi \cdot r$. Therefore, 3 must be smaller than π.

These pictures provide us with additional qualitative information: The path along the square is *much* longer than that along the circle, while the path along the hexagon is only a *little bit* shorter. This means that π must be much closer to 3 than to 4.

Chapter 17

How Random Events Become Calculable Quantities

Chance is something that cannot be calculated out beforehand: even the cleverest mathematician can do no better than to determine the probabilities of winning and losing.

Yet that is only part of the story, since uncertainty becomes smaller and smaller as the number of random influences grows. Think, for example, of a coin that with equal probability—that is, 50%—shows heads or tails. One can toss this coin to make a fair yes–no decision, since the outcome cannot be determined in advance. However, if you toss this coin ten times and count up the number of heads, there will be a significant preference for the "average" values. It is much more likely that the result will be five heads (just under 25%) than one head (less than 0.1%). With even more coin tosses this tendency is even more spectacular, the number of heads hovering with overwhelming odds within a small range of half the number of tosses.

What lies behind this phenomenon is one of the most important limit theorems in probability theory. It is such theorems that describe the transition from the unpredictable to the deterministic. This fact is of much more than theoretical interest. For example, in this connection one may recall that according to quantum mechanics, the world is governed on the nanoscale by random processes. It is only the superposition of unimaginably many random events that grants us the illusion of living in a deterministic world. The same principle is at work in the ability of the television-network prognosticators to tell us the outcome of a presidential election after only a small percentage of the votes have been reported, since one can determine the likely percentages from a small random sample.

Finally, the buyer for a supermarket chain and the planners for public transportation rely on limit theorems: it is extremely unlikely that suddenly every customer is going to experience the need to buy baking powder or that everyone living near the local train station is going to decide on taking the 8:50 to Grand Central Station.

Start Your Own Business

There is a whole zoo of limit theorems. All say that chance diminishes as the number of random influences increases. Suppose you decide to set up a gambling booth at the county fair. To keep calculations simple, we will assume that a customer throws a single die one time. If it comes up a six, he or she wins thirty euros. Otherwise, the payout is zero. Therefore, for each round you will have to pay the customer thirty euros with probability $\frac{1}{6}$, and so you should figure on an average payout of $\frac{1}{6} \cdot 30 = 5$ euros.

Therefore it should cost at least that amount to play, since you don't want to lose money. Let us suppose that you set the entrance fee at seven euros. How much, then, is a customer worth? With probability $\frac{1}{6}$ you will suffer a loss of twenty-three euros (thirty euros paid out minus the seven-euro charge), and with probability $\frac{5}{6}$ you will keep the seven-euro fee. Therefore, a customer brings in on average

$$-\frac{1}{6} \cdot 23 + \frac{5}{6} \cdot 7 = \frac{-23 + 35}{6} = 2$$

euros.

If a good day brings in three hundred customers, you can reckon on an income of $2 \cdot 300 = 600$ euros. And on account of the limit theorems of probability theory, with that number of players you can rely on these six hundred euros with almost one hundred percent certainty. It is extremely unlikely that there will be a large number of lucky players and that less than 550 euros will be in the cash register come evening. Unfortunately, you can also scarcely hope to have more than 650 euros.

The Vanishing of Chance: Some Calculations

In this section we consider a *quantitative example* in connection with limit theorems. First we toss a fair coin ten times. It is to be expected that heads will come up five times on average. The exact values are as follows:

Number of Heads	Probability Thereof
Is exactly 5	24.6%
Is between 4 and 5	54.2%
Is between 3 and 7	77.4%

Now let us toss the coin one hundred times:

Number of Heads	Probability Thereof
Is exactly 50	7.95%
Is between 45 and 55	72.9%
Is between 40 and 60	96.5%

And finally, one thousand times:

Number of Heads	Probability Thereof
Is exactly 500	2.52%
Is between 490 and 510	49.2%
Is between 480 and 520	80.6%
Is between 470 and 530	94.6%

Thus while as expected, it is quite unlikely that exactly five hundred heads will turn up, one can almost certainly count on the deviation from this number being at most 30, which is six percent of the number of tosses.

One can actually observe the disappearance of chance. Suppose our friend Isabella plays the following game. With equal probability she wins or loses one euro. The values of the individual games will be denoted by $x_1, x_2, \ldots$, and so a sequence of games might take the form $1, 1, -1, 1, -1, \ldots$. Adding the first n numbers of the sequence yields the net profit or loss after the nth round. In our example, these values for $n = 1, 2, 3, 4, 5$ are the numbers $1, 2, 1, 2, 1$. And now, to determine the *average* win, the profit or loss after the nth round must be divided by n, yielding $1, 1, \frac{1}{3}, \frac{1}{2}, \frac{1}{5}, \ldots$. Observe that these values are moving in the direction of zero. Figure 1 shows a graph of these average values for a particular sequence of games. At the beginning, good and bad luck alternate, followed by a run of bad luck, and then the graph slowly begins to rise. But in the long run, wins and losses *always* average out, and the average win tends to zero.

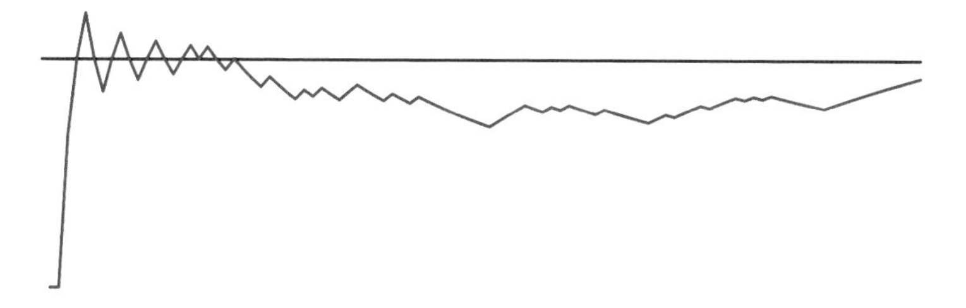

Figure 1. Chance disappears.

Chapter 18

A One-Million-Dollar Prize: How Are the Prime Numbers Distributed?

In this chapter we are again going to discuss prime numbers. The primes have lost nothing of their fascination since they were first investigated well over two thousand years ago. (Recall that prime numbers are the whole numbers that cannot be written as a product of smaller whole numbers; thus, for example, 7 and 19 are prime, while 20 is not.) Even Carl Friedrich Gauss, certainly one of the greatest mathematicians of all time, was drawn into their orbit. He wanted to know how the prime numbers are distributed among the integers. Can one say how many primes there are "far out there"?

Two facts are clear. First, the prime numbers seem to pop up like mushrooms without any particular rule: if you mark the prime numbers among the first hundred positive integers, you will see that they form rather a random pattern. And second, it is clear that a large number has worse chances of being prime than a small one, since for the large one there are more potential divisors.

Gauss proceeded pragmatically, doing what today we would call "experimental mathematics" (and for which today one uses computers). Based on his concrete calculations, he conjectured what today is called the prime number theorem: The proportion of prime numbers less than a given number can be well approximated. For a number with k digits it is almost exactly $0.43/k$. (A more precise formulation appears below.) Thus under 1,000, since $k = 3$, the number is $0.43/3$, that is, about 0.143, or 14.3%; under 1,000,000 the proportion of prime numbers is only $0.43/6$, or 7.2%.

Gauss was long dead when his conjecture was proved to be a mathematical fact. Toward the end of the nineteenth century, independently of each other, the mathematicians Hadamard and de la Vallée Poussin gave rigorous proofs.

But that was far from the end of the story. Since then, there have appeared much-more-refined descriptions of the distribution of prime numbers than that conjectured by Gauss. For the solution of one aspect of this puzzle there has been offered, since the year 2000, a million-dollar prize.

The Prime Number Theorem

To aid in visualizing the growth of prime numbers, Figure 1 shows a graph in which starting from the left, the kth line represents the kth prime at height k. Thus, for example, the fourth line (here shown in blue) represents the fourth prime, namely 7, and therefore the x coordinate of this line begins at 7 and ends at 11, the next prime number.

The prime number theorem now asserts that for large x, the height of the graph, representing the number of primes less than x, is very well approximated by $x/\log x$.

> To understand this expression, one needs to know what is meant by the *natural logarithm* $\log x$ of a number x. It is the number y for which
>
> $$(2.71828\dots)^y$$

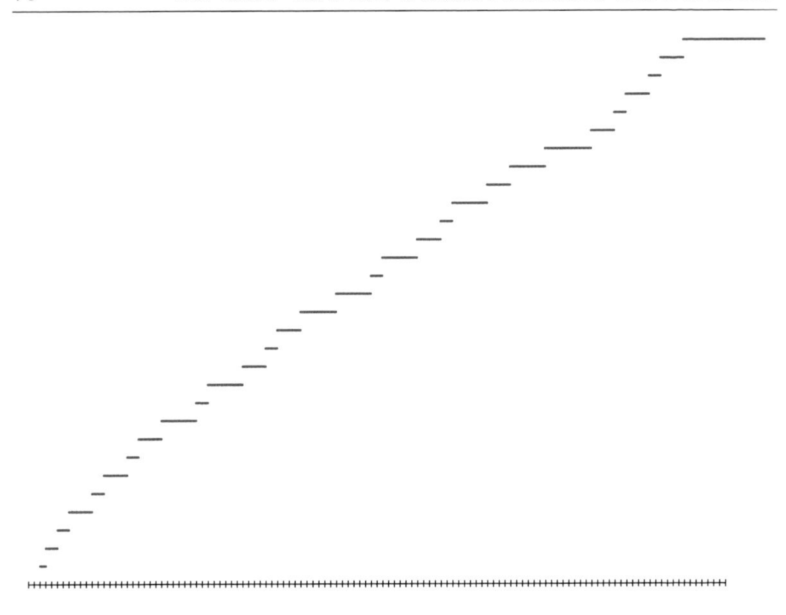

Figure 1. The growth of prime numbers.

is equal to x. In somewhat simplified form, one could say that the logarithm of a k-digit number can be (crudely) approximated by $k/0.43$. (For example, the value of the logarithm of 8,000 is $8.987\ldots$, while $4/0.43 = 9.302\ldots$.)

To appreciate the goodness of the approximation, let us look at a few examples. For $x = 100{,}000{,}000$, there are 5,761,455 prime numbers less than x. The quotient $x/\log x$ differs from this number by 332,774, which represents an error of about six percent. For $x = 10{,}000{,}000{,}000$, there are 344,052,511 primes less than x. The prime number theorem predicts about twenty million primes too few. That's a lot of primes, but the error is just a bit over four percent.

In considering what is meant by "very well approximated" in the above discussion, a more careful analysis reveals that one can do much better with more-complicated closed formulas. The best of these gives the number of primes less than x with remarkable precision.

For example, for $x = 100{,}000{,}000$ the formula is off by only 754, with an error of about one-hundredth of one percent. And for $x = 10{,}000{,}000{,}000$ the formula fails by only 3,104, representing an error of less than one-thousandth of one percent.

However, to prove the precise magnitude of the error, which no one doubts, requires the solution of a famous unsolved problem, the *Riemann hypothesis*. It is for the solution of this problem that the Clay Mathematics Institute has offered a one-million-dollar prize. (For more on this, see http://www.claymath.org.)

Chapter 19

The Five-Dimensional Cake

As a term of disapprobation, for example in reviews of films and books as well as in everyday conversation, one frequently encounters the expression "one-dimensional." What is meant is that the work being disparaged proceeds in a linear fashion, without any ramification. But what do we really mean by one-, two-, three-dimensional? Indeed, what is *dimension*?

Put simply, the dimension of a geometric object is the number of numbers necessary for identifying a point of that object. Take, for example, a line. If one fixes a point P on the line, then every other point can be specified by a single number: simply specify how far the point is from P, using positive numbers for points to the right of P, and negative numbers for those to the left. Therefore, the line is one-dimensional.

Similarly, one determines that the surface of the Earth is two-dimensional. For example, every such point can be identified by its latitude and longitude. In space, one needs three numbers, and if one wishes to specify a point in space and time simultaneously, then one must contend with four numbers, which specify a point in the spacetime of relativity theory.

Mathematicians frequently operate in much higher dimensions. In order to understand what is going on, it generally suffices to consider a two- or three-dimensional example that encompasses the most important aspects of the situation, just as from a two-dimensional photograph one can reconstruct the three-dimensional original. Thus, for example, a five-dimensional space will be simply a set whose points require five numbers for their specification.

This may all sound difficult and abstract, yet there are parallels with our everyday experience. A cake recipe, for example, may be defined by the amount of the various ingredients in grams. If one writes down the amounts of flour, sugar, butter, eggs, and baking powder in the form $(200, 100, 80, 20, 3)$, then this representation contains the most important information. That's it; it's admittedly not particularly exciting, but in fact, there is really nothing very complicated about five dimensions.

A Leap into the Fourth Dimension

Mathematicians are possessed of the same cortical convolutions as their fellow human beings, and therefore they are unable to picture more than three dimensions. Nevertheless, they can work without difficulty with objects of very high dimension. It is important only that one be able to visualize important aspects of a problem in a two- or at most three-dimensional picture. If what is at issue is distances, then a picture should reproduce the separation between points accurately: points that are equidistant in actuality should be equidistant in the representation, and so on. This is in fact not much different from the work of mapmakers, for example, who also represent only the most important aspects of reality in their drawings. No one expects a rendering of every last detail of a bus route; what is important is the information giving, for example, the distance between the stops.

Let us demonstrate this with an example of how mathematicians approach the fourth dimension. We will begin with an exercise in three dimensions: how could one explain the surface of a (three-dimensional) cube to a creature that can imagine only two dimensions? We begin with Figure 1.

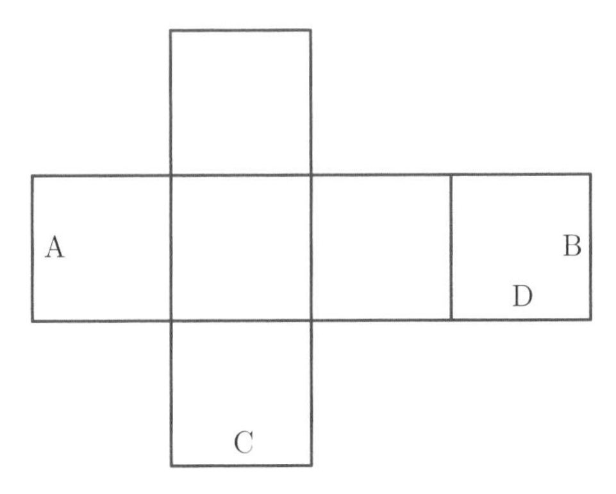

Figure 1. A two-dimensional creature moves along this surface.

The figure represents the usual cutout pattern for a cube. Our two-dimensional creature, let us call him Ferdinand, is placed on it and is given instructions for taking a walk, whereby he has to obey the following conditions:

- You may move about the surface freely, but you may not leave it.
- If you have the illusion that you have crawled off the figure, in fact you have reentered it at another point. More precisely, if you leave the surface at point A, then you reenter at B, while leaving at C is equivalent to reentering at D, and so on.[1]

Ferdinand can thus become accustomed to the surface of the cube, which for us presents no challenge whatsoever. He will observe, for example, that the surface has no boundaries; it never comes to an end. However, the surface is finite; indeed, one could paint it with a finite amount of paint. With some experience, Ferdinand would also internalize some other characteristics of the surface, for example, that to every point there corresponds another point that is as far away as possible. In the three-dimensional world, these are antipodal points.

We shall now repeat all of this in one dimension higher. Now *we* are the three-dimenional creature trying to experience the fourth

[1] The other rules yield the result that the surface is imagined as folded up into a cube; they describe the transitions between adjoining edges.

dimension. We should at least be able to develop a feel for the three-dimensional surface of a four-dimensional cube. To this end, we allow ourselves to be placed in the accompanying picture, showing an object that looks something like a child's jungle gym. We are given a piece of paper on which are presented the "climbing rules":

- You may not leave the jungle gym.
- If you have the illusion of having climbed off, in reality you have reentered at another point. For example, exiting the top is equivalent to reentering at the bottom.
- There are some additional rules, which are not described here.

In this way, one can actually investigate the structure of a geometric construct that is inaccessible to our direct perception.

We close by mentioning that the painter Salvador Dalí (1904–1989) immortalized this hypercube in his 1954 painting *Crucifixion, Corpus Hypercubus*. Perhaps the symbolism is that one can grasp only an idea of a God who is incomprehensible directly.

Chapter 20

One Night Stand

The literary term "composition" is used in mathematics to indicate a new object being formed from two other objects. Thus from two numbers x, y one can form the sum $x + y$ or the product $x \cdot y$. As an example from daily experience one can consider how from the words "foot" and "ball" one can form the word "football." Similarly, meaningful sentences are composed from phrases.

If one is studying such kinds of composition and would like to work with not only two, but three elements, then there is a problem. In adding numbers, for example, given x, y, z, one could first form the sum $x + y$ and then add z, but one could equally well add x to the sum $y + z$. In symbols, the threefold sum could be equally well represented by $(x + y) + z$ as by $x + (y + z)$. If both methods always lead to the same result, one says that the composition is *associative*. This is a very important property, since it can spare one having to deal with a sea of parentheses.

It is well known that addition and multiplication are associative. For example, it is certainly true that $(1 + 2) + 3 = 1 + (2 + 3)$ and $(3 \cdot 4) \cdot 5 = 3 \cdot (4 \cdot 5)$. However, by no means do all important rules of composition have this pleasing property: for example, if one forms from x, y the quotient x/y, that composition is not associative, as can be seen from the fact that $(20/2)/2$ it quite different from $20/(2/2)$, since the first expression is equal to 5, while the second yields 20.

For better or worse, language is not associative, since the way that one associates words can yield quite different results. A number of years ago, a newspaper article appeared with the headline, "Rail Service for Area Dead." Was the local rail service defunct, or had a new service been instituted for transporting local corpses?

Another example comes from the 1958 hit song "Flying Purple People Eater." The following exegesis can be found in the online encyclopedia *Wikipedia*:

> One eternal question about the song is caused by an ambiguity in the English language: is the eponymous creature a one-eyed, one-horned flying purple creature that eats people, a creature that eats one-eyed, one-horned, flying purple people, or somewhere in between? The lyrics clarify matters somewhat: the creature is described as having one eye and one horn, and it comes out of the sky (presumably by flying). However, it is also stated that the creature is a "purple people eater." There are two conclusions to be drawn from this: it is either a purple creature that eats people, or it is a one-eyed, one-horned, flying creature that eats purple people. (The exact color of the creature is therefore open to debate, but most artwork assumes that it, too, is purple.) However, the lyrics also include, "I said Mr. Purple People Eater, what's your line? He said it's eatin' purple people and it sure is fine." Clearly the creature eats purple people, and may or may not be purple itself.

Everyone Wants to Save (at Least Parentheses)

The associative law allows one to "save a couple of parentheses." However, ambiguity due to the absence of parentheses can be severe when associativity is absent. It grows rapidly with the number of elements. If there are three elements, say a, b, c, one must use parentheses to distinguish whether $a \circ b \circ c$ is to be interpreted as $(a \circ b) \circ c$ or $a \circ (b \circ c)$, where the symbol $\circ$ stands for some composition or other.

With four elements, there are four possibilities: Does $a \circ b \circ c \circ d$ mean $(a \circ b) \circ (c \circ d)$ or $\big(a \circ (b \circ c)\big) \circ d$ or $a \circ \big((b \circ c) \circ d\big)$ or $a \circ \big(b \circ (c \circ d)\big)$? The parentheses are necessary, since each of these expressions can mean something different from each of the others.

Without the associative law, then, things become much more complex. Of almost greater significance is the fact that without associativity, one must do without many elementary constructions. For example, if you want to use the abbeviation a^4 for $a \circ a \circ a \circ a$, then you must first make certain that the second expression is uniquely defined. What would the expression a^4 mean if depending on the placement of parentheses, $a \circ a \circ a \circ a$ yielded five different results?

For example, we might consider the composition $a \circ b = a^b$ for natural numbers. It is not associative, since for almost every value of a, it is already important how the parentheses are placed even in calculating something as simple as $a \circ a \circ a$. For $a = 3$, we have $\left(3^3\right)^3 = 729$, but $3^{\left(3^3\right)} = 19\,683$. For larger numbers a the difference is even more spectacular: $\left(9^9\right)^9$ is a number of "only" 77 digits, while $9^{\left(9^9\right)}$ has many millions. Indeed, it is the largest number that one can write with three 9's.

Dog House and House Dog

In addition to associativity there is an additional property of composition that mathematicians consider important: *commutativity*. One says about a composition rule that it is commutative (or that the commutative law holds) if the order of operation is irrelevant, that is, if it is always true that $a \circ b = b \circ a$. Well-known examples of commutative compositions are addition and multiplication. However, the commutative law does not hold for division: 4 divided by 2 is not the same as 2 divided by 4. Also, the composition a^b that we just looked at is not commutative, since, for example, 2^5 is different from 5^2. (Indeed, the numbers can almost *never* be exchanged. To be sure, $2^4 = 4^2$, but otherwise there are no two different natural numbers a and b for which $a^b = b^a$.)

In contrast to associativity, which is almost always assumed, commutativity fails to hold in many important situations. Dealing with objects such as "noncommutative groups" and "noncommutative algebras" can be tough going.

Moreover, in natural language, as was the case with associativity, the commutative law fails to hold. After all, a racehorse is not the same as a horse race, and a dog house is different from a house dog.

And there are still more formal parallels. You may remember from your schooldays that there is a *distributive law* that allows for the removal of parentheses. For example, the expression $a \cdot (b + c)$ is always equivalent to $a \cdot b + a \cdot c$. There are analogies in language: "five- and ten-dollar bills" stands for "five-dollar and ten-dollar bills." And don't forget those hyphens: How did you interpret the title of this chapter: "one night-stand" or "one-night stand"?

Chapter 21

Fly Me to the Moon

Mathematicians generally react with annoyance when in a conversation, their interlocutor, learning of the mathematician's profession, says, "Is there really anything at all new to be discovered in mathematics?" It does not seem quite to have become public knowledge that mathematics is an exciting endeavor, demanding enormous creativity and providing concrete solutions to real-life problems. Therefore, today we are going to look over a mathematician's shoulder in the hope of providing an image makeover. To get an idea of what is at stake, imagine a mountain range covered with a thick, shining coating of ice. If you would like to get from one mountain peak to another of equal height, all you would have to do, in theory, is to point yourself in the right direction and slide down the mountain: the force of gravity would accelerate you, and the energy accumulated in your descent would suffice exactly for the ascent to the neighboring peak.

A similar situation obtains in space travel, though it is much more complex. Just as in the mountain-peak example, there are paths between certain locations in space that can be traveled with virtually no expenditure of energy if one makes clever use of the gravitational pull of the Sun, Moon, and planets. And in fact, long space voyages are planned using such techniques.

Of course, the points in question must first be calculated, and one must figure out the minimal course correction necessary for maintaining the correct direction. The mathematical requirements for such calculations are enormous, and one may safely say that without the development of the theoretical and applied mathematical basis over the past few decades and with the enormous capabilities of today's computers, such calculations would have been impossible to carry out.

This is only one mathematical application among many that we might have mentioned.

"Off-the-Shelf" Mathematics

Over the centuries, mathematics has accumulated an enormous reservoir of methods and results that stand ready to be taken off the shelf and put to use. It is certainly true that most of the results of the last hundred years or so came about because the underlying problem presented a certain fascination to the mathematical mind. Concrete applications were often not considered at all.

However, it frequently happens that problems arise in practical applications that can be solved by adapting mathematical tools that already exist.

A famous example of this from early history is that of the *conic sections*: these are the curves that arise when one slices a cone with a sharp knife, yielding circles, ellipses, parabolas, and hyperbolas. Much was known about these curves already in ancient Greece, and an early standard work on the subject is the treatise *Conics* by Apollonius from about 200 B.C.E.

Seventeen hundred years later, after the collapse of the Byzantine Empire, the mathematical knowledge of ancient Greece began to arouse interest in Central Europe. However, the many translations and transcriptions of mathematical texts over the centuries had led to the introduction of numerous errors. The best mathematicians of the sixteenth and seventeenth centuries set about to restore, as much as possible, the original texts. Thus did the work of Apollonius become known in intellectual circles.

Apollonius's treatise was of great importance to the astronomer Johannes Kepler as he was attempting to bring Copernican cosmology into alignment with existing observational data. Copernicus had written that the planets move in circular orbits about the Sun, but that was not quite true, as was discovered at the beginning of the seventeenth century when astronomical measurements became more precise. It was Kepler's brilliant idea to replace the circular orbits with elliptical ones. And everything that he wanted to know about ellipses was to be found in Apollonius.

There are many such examples. Einstein's theory of relativity (beginning of the twentieth century) is unthinkable without Riemann's differentiable geometry (middle of the nineteenth century). And the mathematics of computer-aided tomography—a technique developed in the 1960s for reconstructing a three-dimensional object from the measurement of the damping of rays sent through this object in various directions—was essentially complete fifty years earlier.

But those are the exceptions. As a rule, the theory that is needed for the successful disposition of a problem that arises in applications must be developed from scratch. This interplay of intellectual attraction and the possibility of solving concrete problems contributes greatly to the attraction of mathematics.

Chapter 22

Using Residues

If you bring home 81 gummi bears (see Figure 1) to divide among your five children and attempt to distribute them evenly among them, every child will get sixteen gummi bears, and one will be left over (which you should surreptitiously eat). Mathematicians describe this phenomenon by saying that 81 modulo 5 is equal to 1. In general, m modulo n is the remainder that results when m is divided by n. Calculations "modulo" play an important role in many branches of mathematics.

Figure 1. 81 modulo 5 equals 1.

Nonmathematicians can also work easily with this technique in a number of special situations. For example, if you want to know what day of the week it will be 39 days from now, intuition tells you, correctly, to calculate 39 modulo 7, which is equal to 4. Therefore, in 39 days, the day of the week will be the same as that 4 days from now. And what time will it be in 50 hours? Easy! Calculate 50 modulo 24, with the result 2, and therefore in 50 hours the clock will indicate two hours later than it is now.

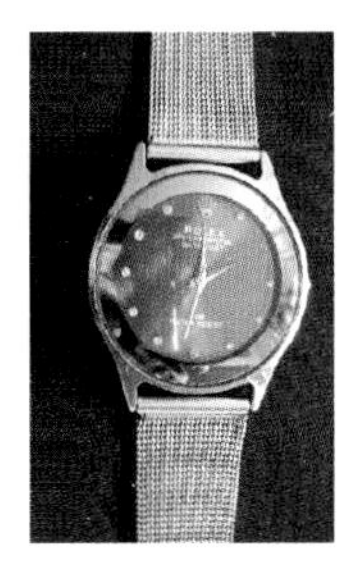

Figure 2. Now... and 50 hours later.

So far, this is nothing particularly noteworthy; we have simply presented a technical term for a well-known method of calculation. For mathematicians, however, there is much more behind all of this, and a number of surprising properties of numbers can best be formulated in terms of modular arithmetic. As an example, suppose that n is a prime number and k a number between 1 and $n-1$ (inclusive). What happens when k is multiplied by itself $n-1$ times? Remarkably, the answer, taken modulo n, is always 1. In the gummi-bear example above, we had $n=5$ (the number of children is prime), and choosing $k=3$, then k $(=3)$ times itself $n-1$ $(=4)$ times yields $3\cdot 3\cdot 3\cdot 3=81$, the number of gummi bears. That 81 modulo 5 is equal to 1 is then a consequence of the general result presented here.

The fact that in the case of a prime number, the above modular calculation always leads to 1 has long been known. It was discovered by the French mathematician Pierre de Fermat in the seventeenth century. Today, it plays an important role in applications in cryptography, where the prime numbers in question have several hundred digits. (More on this can be found in Chapter 23.)

Six Times Six Equals One

The remainders, or "residues" (that is, the numbers $0, 1, \ldots, n-1$), behave just like ordinary numbers with respect to addition and multiplication, provided that one always reduces the result to the appropriate residue modulo n.

> For example, if one is calculating modulo 7, the product of 3 and 5 is equal to 1, since $3 \cdot 5$ modulo 7 is equal to 1. Similarly, 4 plus 6 is equal to 3, since $4+6$, calculated modulo 7, equals 3.

Modular arithmetic thus has many properties in common with ordinary arithmetic. The parallels are particularly striking when n is a prime number. Then each number (other than zero) can be multiplied by some other number to yield 1. As an example, take the number 6, and we will again calculate modulo 7. Taking the successive products $1 \cdot 6$, $2 \cdot 6$, $3 \cdot 6$, $4 \cdot 6$, $5 \cdot 6$, $6 \cdot 6$, we obtain the remainders $6, 5, 4, 3, 2, 1$, and so we see that $6 \cdot 6$ is equal to 1.

This property does not hold for numbers that are not prime. For example, if we choose $n = 12$, then you will search in vain for a number x with the property that $4 \cdot x$ is equal to 1 modulo 12. This is because dividing $4 \cdot x$ by 12 always produces one of the number $0, 4, 8$ as remainder.

This wealth of algebraic properties suggests the mathematical significance of modular arithmetic. For example, the *commutative* property of addition continues to hold; that is, the sum $a+b$ is always equal to $b+a$ under addition modulo some number n.

Chapter 23

Top Secret!

Prime numbers have popped up in this book on a number of occasions. In this chapter we will have something to say about how large prime numbers have revolutionized cryptography, the science of secret codes.

Suppose you have determined two very large prime numbers—call them p and q—that only you know. Here "large" means that they each have several hundred digits. We then compute the product $p \cdot q$, which we shall denote by n.

Figure 1. Classical cryptography: the "Enigma" machine.

Remarkably, the numbers p and q are hidden within the number n like needles in a haystack. In particular, there is no known technique to find the factors p and q, given their product n, in a reasonable amount of time, even if the world's fastest computers were to work for several millennia.

It is this fact that contemporary cryptography exploits. It uses a theorem of number theory that has been known for centuries: one can carry out a series of manipulations using n in such a way that the process can be inverted only if one has knowledge of p and q. Therefore, if your friend Isabella wishes to send you a secret message, you have only to send her the number n and describe a procedure for how her message, after being represented as a large integer, is to be further encoded using the number n. She will then send you the result of this transformation, and now no one other than you can make head or tail of the encrypted message, while you, with knowledge of p and q, can easily decode the message.

What is revolutionary about this process is that it all takes place practically in full public view. Everyone can know the number n, know the procedure for encryption, and see the encrypted message. One therefore speaks of "public key cryptography."

The mathematical aspect of all this, vaguely mentioned above as a procedure for further encoding using the number n, relies on modular arithmetic as described in the previous chapter. Even mathematicians find it surprising that this form of arithmetic is used millions of times every day for sending encrypted information, for example in Internet traffic.

Encryption Using the RSA Procedure

To understand more fully what is meant by public key cryptography, one must know a few terms and basic mathematical facts. In basic outline, the so-called RSA procedure works as follows.[1]

Fundamentals. The basic idea is to use modular arithmetic, as discussed in Chapter 22. One needs to know, for example, why the

[1] This procedure was proposed in 1977 by Rivest, Shamir, and Adleman, whence the name.

equation 211 mod 100 = 11 is correct.[2] And if one has a computer, one may convince oneself that

$$265{,}252{,}859{,}812{,}191{,}058{,}636{,}308{,}480{,}479{,}023 \bmod 1{,}459{,}001 \\ = 897{,}362.$$

Facts. In Chapter 22 we mentioned a surprising fact: if n is a prime number and k an integer between 1 and $n-1$, then it is always the case that

$$k^{n-1} \bmod n = 1.$$

Mathematicians call this fact "Fermat's little theorem."[3] If we multiply both sides of the equation by k, then we obtain

$$k^n \bmod n = k.$$

We are not going to give a proof here, but we will use the result in the discussion that follows.

> Let us give a numerical example as illustration: If $n = 7$ and $k = 3$, then $k^n = 3^7 = 2{,}187$. And indeed, $2{,}187 \bmod 7 = 3$.

We will need a generalization of Fermat's theorem for integers that may not be prime that was first proved by the mathematician Leonhard Euler (1707–1783). To formulate this result, one must know what the term "relatively prime" means: two integers m and n are said to be *relatively prime* if the only number that divides both m and n evenly is 1. Thus, for example, 15 and 32 are relatively prime, while 15 and 12 are not (since they have the common factor 3).

If n is a positive integer, we define $\phi(n)$ (say "fee of n") to be the number of integers between 1 and n (inclusive) that are relatively prime to n. For example, for $n = 22$, we have that

$$1,\ 3,\ 5,\ 7,\ 9,\ 13,\ 15,\ 17,\ 19,\ 21$$

are relatively prime to 22, and so $\phi(22) = 10$.

[2] Here 211 mod 100 = 11 is shorthand for "211 modulo 100 is equal to 11." In what follows we will generally use the more compact formulation.

[3] Fermat's "big" theorem is the much, much more difficult problem of determining whether there can exist an equation of the form $a^n + b^n = c^n$ with nontrivial solutions in integers for $n > 2$. See Chapter 89.

We may now state Euler's result: if k is relatively prime to n, then

$$k^{\phi(n)} \bmod n = 1.$$

As a test, let us take $n = 22$ and $k = 13$. Then

$$k^{\phi(n)} = 13^{10} = 137{,}858{,}491{,}849,$$

and 137,858,491,849 modulo 22 is in fact equal to 1.

(For those who would prefer an example in which one can do the calculations mentally, try $n = 6$ and $k = 5$. Then $\phi(6) = 2$, and $5^2 \bmod 6 = 1$.)

One should note that Fermat's little theorem is a special case of Euler's theorem. That is, if p is a prime number, then p has no common factor with any smaller positive integer other than 1 (by definition of what it means to be prime). Therefore, all the integers $1, 2, \ldots, p-1$ are relatively prime to p, and so $\phi(p) = p - 1$. Thus in this case, Euler's theorem becomes Fermat's little theorem.

The RSA Procedure

To begin, we need to find two large prime numbers p and q and then calculate the product $p \cdot q = n$ (where we remind the reader that "large" means a number of several hundred digits). We will also require two additional numbers k and ℓ such that $k \cdot \ell$ is equal to 1 modulo $\phi(n)$. Now, the only numbers between 1 and n that are not relatively prime to n are multiples of p and q, since p and q are prime, and therefore, $\phi(n) = (p-1) \cdot (q-1)$.

Here is an example: In the case $p = 3$ and $q = 5$, we have $n = 15$. The numbers between 1 and 15 that are relatively prime to 15 are

$$1,\ 2,\ 4,\ 7,\ 8,\ 11,\ 13,\ 14,$$

and thus exactly $8 = (3-1)\cdot(5-1)$ integers. Therefore, $\phi(15) = 8$.

Our preparations are complete. The numbers p, q, and ℓ are to be kept under lock and key, while n and k will be publicized to all interested parties. Now if anyone wishes to send a message, it should first be translated into a string of numbers (for example, using the standard computer ASCII code). Then the string is divided into blocks, say of fifty digits each.

And now the encryption can begin. Suppose Isabella wishes to send her brother Ferdinand an encrypted message. If a block is represented by the number m, say, then she must compute $m^k \bmod n$ (we will call the result r). She can do this, since n and k are public knowledge. She then repeats this for each block and sends Ferdinand the result (that is, the numbers r). Note that anyone who so wishes may read this transmission.

The decryption goes like this: Ferdinand opens the safe in which he has been keeping p, q, and ℓ, and computes $r^\ell \bmod n$. We have $r^\ell = \left(m^k\right)^\ell$, and $k \cdot \ell$ is equal to 1 modulo $\phi(n)$. Therefore, there exists an integer s such that $k \cdot \ell = s \cdot \phi(n) + 1$. And so we have

$$\begin{aligned} r^\ell \bmod n &= m^{k\ell} \bmod n \\ &= m^{s\phi(n)+1} \bmod n \\ &= m \cdot \left(m^{\phi(n)}\right)^s \bmod n. \end{aligned}$$

From Euler's theorem we may conclude that $m^{\phi(n)}$ (and therefore the sth power of this number as well) is equal to 1 modulo n. Altogether, this means that

$$r^\ell \bmod n = m \bmod n = m,$$

where the last equality is true because we have made certain that $m < n$, and thus one can indeed reconstruct m from the publicly transmitted r.

But this reconstruction can be accomplished only by someone with knowledge of $\phi(n)$, that is, of $(p-1) \cdot (q-1)$. Anyone who can determine p and q from n would be able to read the secret message. And that is the reason why the problem of factorization has been given such a great deal of attention in recent years.[4]

Here we look at a concrete example with small numbers (those used in serious applications are much larger). We have chosen the numbers $p = 47$ and $q = 59$, and we publish the product $n = 47 \cdot 59 = 2{,}773$. Now we need to choose k and ℓ. We decide on $k = 17$ and $\ell = 157$. Since $\phi(n)$ is equal to $46 \cdot 58 = 2{,}668$, and since $17 \cdot 157 =$

[4] See, for example, Chapter 43.

2,669 and therefore equal to 1 modulo $\phi(n)$, these choices are indeed suitable ones. The numbers 2,773 and 17 are made public, but 47, 59, and 157 are considered top secret.

Now comes the encryption. Suppose that a message has been converted to the number 1,115. Isabella lets her computer calculate the number $1{,}115^{17}$ mod 2,773. The result is 1,379. She writes this number on a postcard, and the following day it appears in Ferdinand's mailbox. Now he asks his computer to calculate $1{,}379^{157}$ mod 2,773. After a few milliseconds, the result is printed: 1,115. If Isabella's nemesis Mira the Malevolent happens to have secretly copied the contents of the postcard, she will be unable to decipher the secret message.

Chapter 24

Magical Mathematics: Order amidst Chaos

Order amidst chaos. That could be the motto of the mathematical magic trick that I would like to present in this chapter. You will need a deck of playing cards with equal numbers of red and black cards. A standard 52-card deck should be fine for our purposes. As a preparatory move, you should arrange the cards so that colors alternate, as shown in Figure 1.

Figure 1. This is how the cards should be arranged.

And now we let chance intervene three times in this pack of cards. In step one, someone cuts the deck somewhere approximately in the middle. In the second step, someone else shuffles the two halves together. Finally, a third person cuts the pack at a point such that two cards of the same color are separated. See Figure 2.

Figure 2. Cut, shuffle, and cut again...

The two halves are placed one on top of the other and given to you. A naive person would think that these three random processes have resulted in a chaotic mix of the cards, the cards arranged totally at random. And so it appears at first glance. However, a remarkable phenomenon is at work. It turns out that cards 1 and 2 have different colors, as do cards 3 and 4, 5 and 6, and so on. As the magician, you may now place the pack of cards under a cloth and murmur mystical incantations and then pull out pairs of oppositely colored cards as if by magic, even though you are actually just drawing the cards from top to bottom. See Figure 3.

Figure 3. ...and now present the cards a pair at a time.

The mathematics behind this is interesting. The fact that the cards are arranged pairwise with different colors after the three random events can be proved with combinatorial methods. In this connection, mathematicians speak of an *invariant*. The magician Gilbreath, who invented the trick at the beginning of the previous century, seems to have discovered it by trial and error.

A Variant of the Trick

For those who would like to add this trick to their repertoire, here is a variant. Recall that the original goes according to the following outline:

- Prepare the pack of cards (even number of cards of alternating colors).
- Cut the deck, and then shuffle the cards.
- Cut the deck at a point where there are two cards of the same color and reassemble the deck.

Then each pair (cards 1 and 2, cards 3 and 4, etc.) contains two cards of opposite colors.

And now the variant: The deck of cards is prepared just as in the original version, and it is again cut. Warning: This time you must somehow inform yourself as to whether the cards on the bottoms of the two halves are of the same color or different colors. This could be done, for example, as the cards are handed to the shuffler.

The next step is again as previously: the two halves are shuffled together. And that is it: you do not need to cut the pack again.

> The advantage over the first variant is that you don't need to have someone look at the cut deck to see where two cards of the same color reside. Thus no one will get the idea that the colors red and black are more regularly distributed than would be the case in a well-mixed pack of cards.

There are now two cases: Case 1 is that the two bottom cards had different colors. Then no adjustment is necessary. It is guaranteed that every pair—cards 1 and 2, cards 3 and 4, and so on—are of opposite colors. The second case is that the two cards on the bottom were both red or both black. Now things are a bit more complicated. While you are muttering your magic formulas, move the top card to the bottom of the deck. Then again, all pairs will contain one red card and one black. Of course, you don't have to move the top card to the bottom. Instead, your first pair should consist of the top and bottom cards. Thereafter, things proceed as before. Good luck with your magic!

And where is the mathematics in all of this? It guarantees that the trick will always work. It can be *proved* that the cards end up as described here. However, the rather complex theory necessary for the proof is beyond the scope of this book.

Chapter 25

How Does One Approach Genius?

How does one understand an exceptional phenomenon? Carl Friedrich Gauss, who lived from 1777 to 1855, is considered by many to be the greatest mathematician who ever lived. In the period before the introduction of the euro, that he was seen as a cultural asset is evident from his appearance on the ten-mark note together with the graphic representation of some of his accomplishments. For example, there one can see the famous bell curve, an indication of Gauss's contributions to probability theory.

There is scarcely a mathematician alive who could rightly say that he or she truly understands the phenomenon that is Gauss. His publications set the standard of mathematical research for many decades, and it is also noteworthy that he deliberately withheld from publication a number of his results. He did this in part because he believed that his contemporaries were not ready to receive them, and partly because he viewed a number of his discoveries, which today are seen as groundbreaking, as of insufficient interest.

Figure 1. Brocken peak in the Harz Mountains.

Thus, for example, he held, correctly, that the time was not yet ripe for non-Euclidean geometry. Mathematicians (and also philosophers such as Kant) had over the millennia become accustomed to the belief that there could exist only one type of geometry, namely that presented by Euclid almost $2,500$ years earlier, in which a triangle has an angle sum of 180°, every line has a unique parallel through a given point outside the line, and so on.

Gauss, however, realized that Euclid's geometry represented only one of many possible geometries. In 1821, he verified through an experiment in which he measured the angles of a large triangle that in our world, Euclidean geometry in fact reigns, at least within experimental error. The vertices of the triangle that he measured were three mountain peaks: Brocken, in the Harz Mountains (see Figure 1); Inselsberg, in the Thuringian Forest; and Hoher Hagen, not far from Göttingen.

It was only years later that it was realized that non-Euclidean variants of geometry could be used to describe nature, such as in the general theory of relativity.[1]

It would not be doing Gauss justice to view him solely as a mathematician. Equally renowned are his accomplishments in physics (on magnetism) and in astronomy, where he used completely new mathematical methods to calculate celestial orbits. His prediction of the

[1]More on the topic of non-Euclidean geometry can be found in Chapter 80.

position of the asteroid Ceres made his name known in professional circles when he was still a young man.

His significance is reflected in the fact that his name is frequently encountered even today. Recently, one of the most prestigious mathematical prizes in the world was named in his honor, and the most important event held by the German Mathematical Society is—of course—the Gauss lecture, which takes place every semester at a different university.

The Seventeen-gon

At the tender age of 17, Gauss discovered a remarkable relationship between number theory and geometry. It concerns the construction of polygons having n sides and all of whose angles are equal.[2] By definition, such constructions must be carried out with straightedge and compass alone.

Those who studied geometry in high school will perhaps recall that for $n = 3$ such a construction is easy: for an equilateral triangle, one has only to draw a line segment, set the compass to this length, and draw two circles of this radius with centers at the endpoints of the segment. Either of the two points where the circles intersect can be used as the missing third vertex of the triangle. The case $n = 4$, that is, a square, is not difficult, since there is a simple procedure for constructing a right angle. But what happens for other values of n?

It had been known since antiquity that the regular pentagon ($n = 5$) and hexagon ($n = 6$) are constructible. May one conclude that all regular n-gons are constructible? No! Today, one knows exactly the values of n for which such a construction with straightedge and compass is possible. To find out which they are, one begins with the prime numbers that can be written as a power of 2 plus 1. Such primes are called *Fermat primes*. The largest such prime known today is 65,537, and simpler examples are $5 = 2^2 + 1$ and $17 = 2^4 + 1$. If n is a Fermat prime or the product of distinct such primes (which may additionally be multiplied by an arbitrary power of 2), then the corresponding regular n-gon is constructible, and these are the

[2]Mathematicians call these *regular n-gons*.

only values of n that work. For example, since 7 cannot be written as $2^k + 1$, it is not a Fermat prime, and therefore a regular 7-gon cannot be constructed with straightedge and compass. (Of course one can draw an approximation to such a polygon, but that is not the mathematical point of interest.)

Figure 2. A 17-gon.

Trumping the Teacher

As with other great figures in cultural history, many anecdotes about Gauss have come down to us. They serve to emphasize some aspect of his character, even if they may be not entirely true or even completely apocryphal.

Here is the best-known anecdote (presented in the hope that it is new to at least a few readers): A few weeks after the beginning of first grade, Gauss's teacher decided to keep the class busy by having each student add up the first hundred numbers, that is, to calculate the sum $1 + 2 + \cdots + 100$.

A minute later, Gauss announced to the teacher that he was finished, and presented the correct answer, 5,050. Instead of carrying out the tedious calculation like all the other pupils, he reassembled the sum in his head thus: instead of $1 + 2 + \cdots + 100$, he calculated

$$(1 + 100) + (2 + 99) + \cdots + (50 + 51).$$

The advantage is clear, since each of the summands in parentheses has the same value, namely 101. It remained, then, to multiply 101 by the number of summands (there are 50), from which the result $50 \cdot 101 = 5{,}050$ easily follows.

As in other areas of mathematics, and indeed in "real life" as well, this example shows that it is often the point of view from which a problem is approached that determines whether it is easy or difficult.

Chapter 26

On Semitones and Twelfth Roots

There is a persistent belief that mathematicians have a particular affinity for music. However, such was not borne out in a quick poll taken among mathematicians at a recent conference, which suggested that mathematicians are no more musically inclined than, say, doctors or lawyers. However, what is true is that there is a close relationship between the two subjects, mathematics and music.

Over 2,500 years ago, Pythagoras understood that two notes played simultaneously on stretched strings have a particularly pleasing sound when their lengths obey a simple mathematical relationship. For example, if one string is twice as long as the other (and they are otherwise identical as to material and tension), the shorter string sounds exactly one octave higher than the longer one (and as we now know, the frequency of the higher tone—the number of vibrations per second—is exactly twice that of the lower tone). In the case of a perfect fifth, the relationship is 2 to 3. The Pythagoreans built an entire musical scale on this principle, but it remains hidden to us what brings about the connection—to be found in all cultures—between simple mathematical relationships and the pleasures of music.

Unfortunately, the Pythagorean scale and related musical systems possess a decided drawback: when one attempts to modulate from one key to another, thereby setting a different note as the fundamental tone of the scale, the mathematical relationships in the new scale are no longer quite the same as in the old one.

Out of this difficulty was born the idea of dividing the octave democratically into twelve equal parts. From one semitone to the next, the frequency increases by the twelfth root of 2, thus by a factor of 1.059463094.... It was over three hundred years ago that the *equal-tempered* scale was developed, and in his *Well-Tempered Clavier*,[1] Johann Sebastian Bach (1685–1750) demonstrated—by presenting a collection of pieces (preludes and fugues) written in each of the twenty-four major and minor keys—that one could play in any key without having to retune the instrument.

This development by no means exhausted the possibilities of relating mathematics to music. In the twentieth century, many composers used a variety of mathematical relationships in their compositions, from the method of tuning to the large-scale compositional form. For example, the composer Iannis Xenakis (1922–2001) used probabilistic methods, game theory, and group theory as organizing principles in his compositions.

However, no matter how high a value is placed on mathematics, it will never be possible to understand our enjoyment of a Schubert sonata or our favorite pop song in terms of a mathematical formula.

Pythagorean versus Chromatic

Why did the twelfth root of 2 pop up in our discussion of equal temperament? Suppose that the octave is to be divided into n parts, where n is any positive integer. A guitar builder would then have to supply n frets up to the middle of the fingerboard, where the last fret would be exactly in the middle to sound the octave. See Figure 1. If all the musical intervals are to be of the same size, then the frequency relationship between the note on the first fret and that on the open string must be the same as that between the second fret and the first, and so on.

[1]The term *well-tempered* refers to any one of a number of tunings in which the semitones are approximately equal. We will reserve *equal-tempered* for a tuning in which the ratio between semitones is exactly the twelfth root of 2.

If this frequency relationship is denoted by x, then the calculations are straightforward. If two notes are played simultaneously (say on two identically tuned guitars) that are k semitones apart, then the frequency relationship is x^k. In particular, the nth note, which should sound the octave, must have frequency twice that of the lower tone, and thus satisfy $x^n = 2$. In equal temperament, we have $n = 12$, which leads to the equation $x^{12} = 2$, or $x = \sqrt[12]{2} = 1.0594\ldots$.

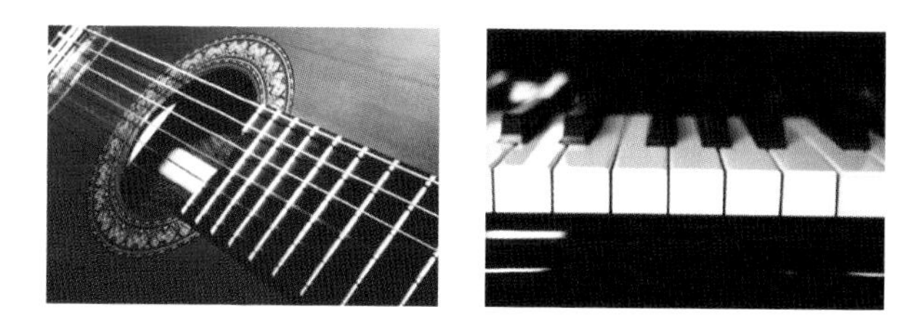

Figure 1. Instruments with equal temperament.

Thus from C# to C the frequency ratio is $1.059\ldots$, and the same ratio holds for D to C#, and so on. One can also calculate the frequency ratio between any pair of notes. For example, the ratio from D to C is calculated as

$$\text{D to C} = \text{D to C\# times C\# to C} = 1.0594\ldots \cdot 1.0594\ldots = 1.12246\ldots.$$

The following table shows the frequency relationships for the phythagorean and equal-tempered C-major scales:

Note	Pythagorean	Equal-Tempered
C	1	1
D	1.12500	1.12246
E	1.26563	1.25992
F	1.33333	1.33484
G	1.50000	1.49831
A	1.68750	1.68179
B	1.89844	1.88775
C	2	2

The frequency ratios are almost identical, and untrained ears will hardly detect a difference. In popular music, equal temperament is almost universal, but when music is played on period instruments from earlier times, the performers frequently attempt to make the music sound as it did at the time of composition.

Chapter 27

Why Am I Always Standing in the Wrong Line?

In this chapter the topic is again psychology. Have you ever had the feeling that the other lines at the supermarket, tollbooth, or post office always move faster than the one you are standing in? It will please you to know that everyone feels that way, and the reason why this is so is easily given.

Imagine that there are five lines—let us say at the post office—of approximately the same length and you have to choose one of them to stand in. Then the probability that you will randomly choose the one that actually happens to be the fastest is $\frac{1}{5}$, or 20%. To put it another way, with 80% probability you will again be waiting in the wrong line. And if you find yourself frequently in such situations, you will undoubtedly develop the impression that fate is not kind to you.

That expectation and reality are often far apart, due to an apparently insufficient mathematical intuition that is part of our evolutionary inheritance, is a frequent topic in this book. For example, exponential growth is not easy to grasp, and the correct solution of the Monty Hall problem (Chapter 14) is intuitively almost impossible for many people.

To the problem presented at the beginning of this chapter we should add that the concept of waiting in lines has been under serious investigation for a long time: "queuing theory" is one of the classical subfields of probability theory.

Queuing theory has many applications. Once one has completely understood a queuing problem, that understanding can be applied to such diverse situations as optimal timing of traffic lights and the transmission of data packets to a junction point of an Internet connection.

Queues

Queuing theory, as we have said, is a branch of probability theory. To describe a typical result, imagine a business that serves the general public: customers arrive, are served, and then leave. It could be a restaurant, a locksmith—you get the idea. We could even think of visitors to a museum or to some tourist attraction as customers.

We now make the following assumptions:

- The customers arrive at random and individually. Here "random" means that one cannot say precisely when the next customer will arrive, only that there is a certain average interval between arrivals (the technical terminology is "exponentially distributed arrival times"). There is also no provision for customers arriving in groups.[1] However, what we can say is that there is a known expectation: *on average*, a customer arrives every K seconds.
- When a customer enters the "store," he or she is served immediately (there are sufficiently many employees). As for

[1]That puts a crimp on our example of tourists. We would have to assume that they are all traveling alone.

the length of time that a customer remains, just as for the arrival times it cannot be predicted precisely, but there is a known expected value, which we will denote here by L: the average number of seconds that it takes to serve a customer.

Depending on the situation, these conditions will correspond more or less to reality. They certainly hold for a large restaurant with ample staff, provided that there are quite a few empty tables. It would seem as well that our conditions would hold for visitors to a historic cathedral.

The parameters K and L are independent, and they depend on the situation. For the cathedral, say, a small value of K means many visitors, while for a large value, they are few and far between. And L is here a measure of the attractiveness of the site. For a small value of L, a tourist does not spend much time on average viewing the cathedral, while a larger value indicates a longer visit (think of Saint Peter's Basilica, in the Vatican).

The problem is to predict the number of customers present at any one time. Qualitatively, the situation is clear: a large K and small L indicate that on average there will be few customers present at any given time. However, we would like to know more precisely what to expect: How many seats for waiting customers should the locksmith provide? How many waiters should the restaurant engage? Such predictions are possible using probability theory.

Here is the result: Let λ denote the quotient L/K, and thus λ represents the number of customers present in the store on average at any given time. Then the probability that at a particular moment exactly k customers are present is given by the number

$$\frac{\lambda^k}{k!}e^{-\lambda},$$

where $k!$ (read "k factorial") is the abbreviation for the product $1 \cdot 2 \cdots k$, and $e = 2.718\ldots$ is Euler's number, the base of the natural logarithm.[2]

Here is an example: Suppose $K = 60$ and $L = 120$. That is, on average, a customer arrives every sixty seconds and stays an average of

[2]More on this can be found in Chapter 42.

two minutes. Therefore, $\lambda = 2$, and one can calculate the probability that at any given time there will be exactly k customers in the store. A few such values are presented in the following table:

k	0	1	2	3	4	5
Probability	0.135	0.271	0.271	0.180	0.090	0.036

Thus if our locksmith provides seating for four customers, it will seldom happen that a customer has to stand: the probability of at most four customers is $0.135 + 0.271 + 0.271 + 0.180 + 0.090 = 0.947$, and so the probability of five or more is $1 - 0.947 = 0.053$, or somewhat over 5%.

Chapter 28

Zero: An Undeservedly Underrated Number

Numbers are abstractions. A set of five pears and a set of five apples have a property in common that can be observed in certain other sets—namely those with exactly five elements. Thus arises the concept of "five," and it has turned out to be practical to introduce a symbol for it. Such abstraction occurs in all human cultures, and even toddlers can operate with such simple numbers.

But what about the number zero? It is certainly not remarkable that some sets have no elements in them, yet it was centuries before the insight was reached that it would be useful to have a symbol for the concept of zero. For example, in the Roman system of notation there was no zero, and in fact, Roman numerals are completely unsuitable for performing arithmetic operations. It was only with the introduction of zero and a numerical notation using place values that large numbers could be easily represented and arithmetic operations conveniently executed.

Those who have mastered their multiplication tables and know how to add single-digit numbers can carry out all the arithmetic operations on large numbers without difficulty. In all of this, zero plays a role of fundamental importance. For example, in the notation of the number 702, it is used to express the fact that the number has no tens (only 7 hundreds and 2 ones). And the more zeros a number has at the end, the larger the values of the numbers that precede it: the 1 in 1,000 counts for much more than the 1 in 10.

In the place-value system in India, the zero was originally indicated only by a special mark, indicating that in that place there was no entry. (That led to fewer errors than if nothing at all were written.) In his very readable book *The Nothing That Is: A Natural History of Zero*,[1] Robert Kaplan writes that at that time in India, zero "was no more a number than the comma is a letter." It was only at the beginning of the sixteenth century that zero became established as a full-fledged number.

For mathematicians, zero's role goes far beyond its job in the representation of numbers. Indeed, it is one of the most important numbers of all. This importance rests on the fact that the addition of zero to another number does not change the result. One speaks of zero as a *neutral element* or the *additive identity*. Among the numbers it can be seen as the center, standing as a fulcrum between the positive and negative numbers.

Even now, the role of zero is not precisely established. In the year 2100, long before New Year's Eve there will be discussion about when the twenty-second century actually begins. It all depends on whether counting of years is taken to begin with 0 or with 1.

How Does One Find the Great Unknown?

We shall use a simple problem involving addition to demonstrate how the properties of zero are used in calculation. We shall not go outside the set of integers, that is, the numbers

$$\ldots,\ -2,\ -1,\ 0,\ 1,\ 2,\ 3,\ \ldots.$$

As a first step you should convince yourself that as mentioned above, zero does not change the result of an addition. Regardless of the value of the number y, we always have $y + 0 = 0$. Then observe that we can "always get back to zero." This means that for every number y one can find a number w such that $y + w = 0$. For example, one can choose $w = -5$ if $y = 5$ is given, and for $y = -13$, we have $w = 13$

[1] Oxford University Press, 2000.

as the number we need. Usually, we write $-y$ for this number w and call w the *additive inverse of* y. Note that we have established that $-(-13) = 13$, which is justification for the rule "minus a minus is a plus."

We are now prepared to solve algebraic equations. Suppose we are looking for a number x such that the equation

$$x + 13 = 4{,}299$$

is true. The unknown number x can be found as follows: Simply add -13, that is, the additive inverse of 13, to both sides of the equation. The original equation is thereby transformed into

$$(x + 13) + (-13) = 4{,}299 + (-13).$$

The left-hand side can be transformed to $x + \big(13 + (-13)\big)$, which is possible because addition obeys the associative law.[2] For the sum $\big(13+(-13)\big)$ we may substitute 0 (which is why we chose the additive inverse!), and instead of $x + 0$, we may write simply x (since 0 is the additive identity, as discussed above). Putting it all together, we have so far the equation $x = 4{,}299 + (-13)$, where it is more common to write simply $x = 4{,}299 - 13$. Now all we need is elementary-school mathematics to see that x is revealed as 4,286.

This all seems a bit complicated for so simple a result. Even mathematicians simply subtract 13 from both sides of the equation $x + 13 = 4{,}299$. However, we wanted to make it clear just this once that it is the crucial role played by the number zero that makes it possible to solve such equations.

---

[2] See Chapter 20.

Chapter 29

I Love to Count!

Combinatorics is an old and honorable branch of mathematics, where it plays an important role in many areas. Its primary task is to count the number of ways that something can happen or be arranged, and it is common for very large numbers to occur. For example, in how many ways can you choose six numbers out of forty-nine in next week's lottery?

Let us imagine an urn that contains 49 balls, numbered 1 to 49 and well mixed. You reach into the urn six times and withdraw a ball, and those are your six lottery numbers.

How many possibilities are there? On the first draw, there are forty-nine, but on the second, only forty-eight, and then forty-seven, and so on. Altogether, there are $49 \cdot 48 \cdot 47 \cdot 46 \cdot 45 \cdot 44 = 10{,}068{,}347{,}520$ ways of drawing six balls. But wait! All these possibilities do not lead to different lottery selections. If six particular balls have been chosen, then all other sequences of drawings are equivalent if they result in the same six numbers, chosen perhaps in a different order. For example, the drawing sequence $2, 3, 34, 23, 13, 19$ results in the same lottery selection as $23, 2, 34, 3, 13, 19$. Now, six balls can be drawn in $6 \cdot 5 \cdot 4 \cdot 3 \cdot 2 \cdot 1 = 720$ different ways; namely, for the first draw there are six possibilities, five for the second, and so on. Therefore, to determine the actual number of different lottery choices, we must divide 10,068,347,520 by 720, which gives us the number 13,983,816 that appeared in Chapter 1.

If you can count, you can determine probabilities. Since only one of 13,983,816 possible lottery bets can win the grand prize, the probability of choosing the winning numbers is 1/13,983,816, which, alas, is depressingly small.

The Four Fundamental Problems of Counting

In counting, we are always considering the number of certain choices: we are looking for k elements out of a total of n elements. Before we go any further, we must make two basic decisions: (1) In our selection, does the order matter? (2) Can an element be chosen more than once?

There are two ways to answer each of these two questions, leading to four cases to distinguish:

Case 1: Order matters, multiple selection possible. As an example, think about the number of possible four-letter "words" (we are allowing here all possible arrangements of four letters, including such nonsense words as RTGH). The order is important, since TOOT is not the same as OTTO, and multiple selection is possible, since we would like to include such words as OTTO and LILT, which have repeated letters.

The number of such words is easily determined. In each of the k selections, one has n possibilities, and thus $n \cdot n \cdots n = n^k$ altogether. Since there are $n = 26$ letters in the alphabet and we are considering combinations of $k = 4$ letters, there are

$$26^4 = 456{,}976$$

four-letter "words" in all.

Here is another example: If one selects four digits from the set $0, 1, \ldots, 9$, one obtains a four-digit number.[1] It is clear that order matters and that the same number can occur more than once. We

[1] Of course, if zero is the first number selected, then we get a number such as 0,233 or 0,003. We can either decide to consider such "nonsense" numbers, or else realize that what we are counting is the number of integers with four *or fewer* digits.

have $n = 10$ and $k = 4$, and so there are $10^4 = 10{,}000$ possibilities. (This is the number of PINs for ATMs, which is not such a large number when you think about how many people use automatic teller machines. There is probably someone in your town with the same PIN as yours.)[2]

A variant of this counting problem becomes important when different sets arise in the various selection steps. For example, how many ways can you choose your dinner at a restaurant if there are 5 appetizers, 7 main courses, and 3 desserts? The answer: simply form the product $5 \cdot 7 \cdot 3 = 105$. The reasoning is the same as in the previous example.

Case 2: Order matters, multiple selection not allowed. A typical example (with $n = 20$ and $k = 11$) is the selection of the first-string soccer team from the twenty students on the squad.

Here the order is important, since if the goalie were to trade places with the center forward, it would be a different team. And that multiple selection is prohibited is also clear: the same person cannot be both the goalie and the center forward.

There are analogous problems in selecting the officers of a club: president, vice president, secretary, treasurer. The order is important, since if Isabella were the president and Ferdinand the secretary, that would make for a different officer slate from one in which Ferdinand was the president and Isabella the secretary. And again, we are not going to allow one person to hold more than one office, which means that multiple selection is not allowed.

In this case as well the computations are easy: For the first choice, one has n possibilities, then $n - 1$ (since the first object chosen is no longer available), then $n - 2$, and so on until k choices have been made. The total number of possibilities is then the product of the number of possible choices in each step, that is, the number

$$n \cdot (n-1) \cdots (n-k+1).$$

(Note that the fact that the last factor is $n - k + 1$ tells us that there are in fact k factors in this product.)

[2]A rigorous proof that two people have to share the same PIN relies on the *pigeonhole principle*, which will be discussed in Chapter 62.

For our soccer team, this means that we can choose

$$20 \cdot 19 \cdots (20 - 11 + 1) = 20 \cdot 19 \cdots 10 = 6{,}704{,}425{,}728{,}000,$$

that is, over six trillion possible teams.

And for our club, if eight members are candidates for the four offices, then there are theoretically

$$8 \cdot 7 \cdot 6 \cdot 5 = 1{,}680$$

different possible officer slates.

Case 3: Order does not matter, multiple selection not allowed. This is certainly the situation that obtains most frequently, and in fact, it appeared above, where we were concerned with the special case of choosing six numbers out of forty-nine possibilities. There are many other examples from which we might choose:

- How many poker hands are there (choose 5 from 52)?
- How many handshakes are there if n people greet one another at a gathering? It is certainly the number of ways of selecting two people from a set of n. This, then, is the case $k = 2$.
- You have $n = 8$ unread books and you wish to take four of them with you on vacation. How many selections are possible?

The path to a solution was shown for the lottery in the present chapter, and for general n and k, it looks like this:

$$\frac{n \cdot (n-1) \cdots (n-k+1)}{1 \cdot 2 \cdots k}.$$

This expression occurs so often in mathematics that it has acquired its own symbol:

$$\binom{n}{k} = \frac{n \cdot (n-1) \cdots (n-k+1)}{1 \cdot 2 \cdots k}.$$

This symbol is read "n choose k" and is called the *binomial coefficient*.

We can now provide numbers for our examples: There are 2,598,960 different poker hands, and if twenty people meet, there are $\binom{20}{2} = \frac{20 \cdot 19}{1 \cdot 2} = 190$ handshakes.

Case 4: Order does not matter, multiple selection allowed. This case arises seldom. For an example, imagine k balls that are to be distributed among n boxes (see Figure 1). "Selecting" from the n numbers now means deciding on the box in which the next ball is to be housed. Any of the boxes can contain more than one ball, and that means that multiple selection is allowed. And it doesn't matter whether first a ball was placed in box 2 and then one in box 4 or vice versa, and so order does not matter. Finding the number of possibilities is a bit tricky. Here is the result:

$$\binom{n+k-1}{k}.$$

Thus there are

$$\binom{5+2-1}{2} = \binom{6}{2} = \frac{6 \cdot 5}{1 \cdot 2} = 15$$

ways of placing two balls in five boxes, while six balls can be placed in ten boxes in

$$\binom{10+6-1}{6} = \binom{15}{6} = \frac{15 \cdot 14 \cdot 13 \cdot 12 \cdot 11 \cdot 10}{6!} = 5{,}005$$

different ways.

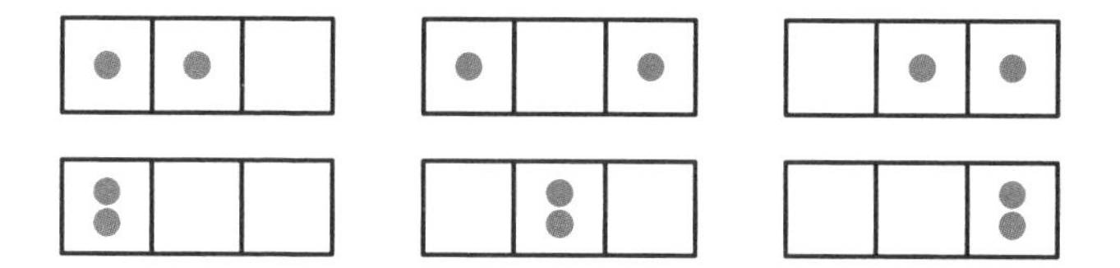

Figure 1. Two balls can be placed in three boxes in $\binom{3+2-1}{2} = \binom{4}{2} = 6$ ways.

Finally, we should note that this problem, which at first glance seems rather esoterically academic, becomes important when one considers, for example, the physics of elementary particles and the number of ways that k electrons can be distributed among n shells.

Chapter 30

Genius Autodidact: The Indian Mathematician Ramanujan

Is there a direct path to mathematical truth? A way that leads to insight without having to devote oneself to painstaking techniques and labor over detailed proofs? It seems to be possible in exceptional cases, and certainly the best-known example is that of the Indian mathematician Srinivasa Ramanujan (1887–1920), whose dramatic life will be briefly recounted here (see Figure 1).

Figure 1. S. Ramanujan and G. H. Hardy.

Ramanujan grew up in poor southern India. He taught himself the fundamentals of mathematics by working through a collection of formulas that came by chance into his hands. Without outside assistance, he discovered notable results in number theory, some of which were known to European specialists, but which for the most part were new. Since he had no university education, he was unable to obtain a position that was suited to his abilities. But somehow he got by, and he devoted every free minute—until overcome by physical and mental exhaustion—to the search for mathematical knowledge.

It was only by a series of lucky chances that he landed at the renowned University of Cambridge. He had begun to write to a number of European mathematicians, one of whom recognized the deep truths hidden among pages of formulas. In Cambridge he had several extremely productive years working with leading specialists. But from overwork and the inability to adapt to the foreign climate and other circumstances he became ill and returned to India in 1919, where he died the next year.

Ramanujan's direct access to truth will always remain shrouded in mystery, but his life story is noteworthy for other reasons. For example, one may speculate how many Ramanujans throughout the world remain undiscovered because their educational development depends on where they were born.

Would Ramanujan Have Had a Better Chance Today?

It was fortunate for the history of mathematics that the English mathematician G. H. Hardy (1877–1947) recognized that the confused letter from India that he had received must have been written by a genius. Other prominent mathematicians had received letters from Ramanujan as well, but they apparently did not take the trouble to look below the surface.

That could easily happen today as well. University mathematicians not infrequently receive letters or e-mails with proofs of famous problems that a glance reveals to be erroneous or something that has long been known. Especially popular are new "proofs" of Fermat's famous last theorem, the squaring of the circle, and the Goldbach conjecture.[1] And every time, without fail, they contain an elementary error. However, the error can be well hidden, and it always takes time and energy to convince the author that he has in fact not given a rigorous proof. And if one simply declines to answer, then one must put up with a barrage of invective: "It is a black mark on you and your university that you are unable or unwilling to appreciate the significance of this important work." Therefore, many institutions, for example the Académie Française, have instituted a policy of simply ignoring such letters.

However, below the Fermat/circle-squaring/Goldbach level, one sometimes receives very interesting ideas from the nonprofessional mathematical community. No Ramanujans have appeared in recent decades, but now and again one is astounded by the original ideas that can be hit upon by those without a formal education.

We close with a quotation from Ramanujan: "An equation means nothing to me unless it expresses a thought of God."

[1]See Chapters 89, 33, and 49.

Chapter 31

I Hate Mathematics Because...

It is an open secret that the majority of my contemporaries have very unpleasant memories of their elementary and secondary mathematical instruction. Children begin school full of enthusiasm. They love to count and calculate, and they can't wait until they will finally be able to count to one hundred. But somewhere between grades 7 and 9 the enthusiasm wanes, the attitude toward mathematics shifts, and thereafter, only a small minority find mathematics an engaging subject.

The reasons for this state of affairs are certainly manifold. One is perhaps that a student has to master a body of rather dry fundamentals before advancing to the interesting stuff. Such is the case in other realms of human endeavor as well: without vocabulary and grammar, no *À la recherche du temps perdu*; without mastery of the C-sharp-minor scale, no *Moonlight Sonata*. However, the danger seems to be especially great in mathematics that a student will find him- or herself mired in technicalities, like a piano student who has been given too many technical exercises to practice and not enough opportunity to make music.

It is also difficult to see at first glance why a study of mathematics —beyond elementary arithmetic—would make one better equipped to solve the types of problems that might help to make this world a better place. The satirical journal *Titanic* once gave a nice example of the supposed disengagement of mathematics from real-world problems: If half a chicken can lay half an egg in half a day, how many eggs can six chickens lay in four days?

You, dear reader, who are perusing this chapter of your own free will, are not in all likelihood among the radical despisers of mathematics. Nonetheless, it would be interesting to find out why so many of your fellow citizens are possessed of a virulent antipathy toward the subject. Suggestions on how to change the situation would be most welcome.

Mathematics in Retreat: An Afterword

The above newspaper column elicited an unusual volume of reader response. The range of opinion is certainly not statistically predictive of the society at large, since those who chose to write are certainly not representative of the general population. Nevertheless, it is noteworthy that two opinions appeared with great frequency:

- Mathematics is disliked because instruction is too advanced. Proofs and the logical structure of mathematics are introduced much too soon and are made the focus of attention. The majority of students are completely befuddled. Quite often, such recollections are accompanied by bitter memories of cynical comments by the mathematics teacher on the class's level of performance.
- It is generally never made clear what the whole thing is good for. A number of readers indicated that the teacher made no connection between mathematics and everyday life. In the best cases, former students have memories of a pleasant intellectual game.

I am perhaps being too optimistic, but there appears to be a tendency toward a more positive view of mathematics. One sees it now and then in advertisements, and indeed not as a mere decorative element (difficult! demanding!), but as an indicator of an intellectual ambience. And it is a long time since one has heard a politician or media celebrity boast about how bad he or she was in mathematics. There are many people working in schools and universities to see that this trend continues.

And this is what we get:

Four out of Three Germans Can't Do Simple Arithmetic

(from an advertisement for a private school)

...and one can imagine the answers on the PISA[1] questionnaire:

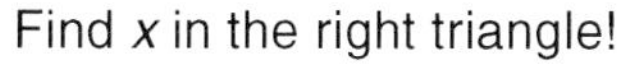

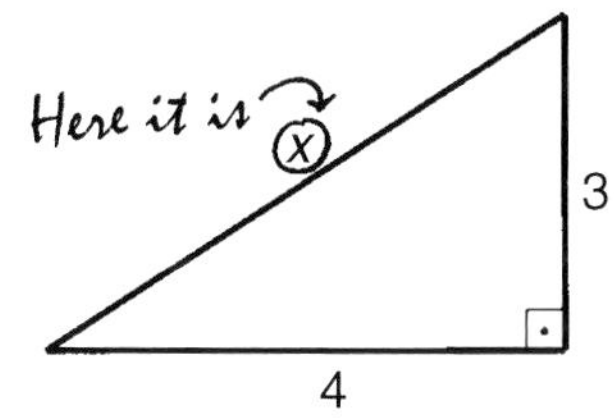

The cartoonist Ulli Stein has attacked this subject and illustrated the following two situations in his book on PISA:

Teacher to class: It's enough to drive one to despair. A full eighty percent of you still don't understand it.
Student, offended: But there aren't even that many of us!

Waiter at pizza parlor: Shall I cut your pizza into four slices or eight?
Customer: Make it four; I couldn't possibly eat eight.

[1]Program for International Student Assessment.

Chapter 32

The Traveling Salesman: A Modern Odyssey

A commercial enterprise has branches in a number of German cities. A representative of the firm is to make a trip, in a company car, to introduce a new product to all the branch offices. How should he plan his route? Since the company is paying his expenses, they would like the salesman to visit each city exactly once and cover the least possible distance. The problem of finding such an optimal route is the famous (among mathematicians) *traveling salesman problem.* The name suggests, falsely, that the problem is confined to optimal highway routes. Such is not the case, and the issues raised by the problem are applicable to any number of planning problems, such as optimizing the path that a computer-driven drill should follow in drilling holes for a circuit board.

Naively, one might think that the solution is easy, since there are finitely many paths, and one can measure them all and choose the shortest one. Theoretically, that is true. However, the number of possible trips is so huge that there is no practical way of carrying out such a plan. Although for real-life traveling salespersons and most other important planning problems in real-world applications there are procedures for finding optimal (or near-optimal) routes in a reasonable amount of time, a fundamental question remains: how difficult in fact is this problem? Is it difficult because in the history of mathematics no one has been clever enough to figure out an efficient algorithm, or is the explosive growth in the number of possible routes as the number of cities grows simply so great that the problem is *inherently* difficult and therefore unsolvable within a reasonable amount of time?

While it may be of little interest to the world's traveling salesmen, I will tell you that the difficulty in the problem is inherent to the extent that it belongs to a class of equally difficult problems, including those on which the security of encryption systems depends. And therefore a million-dollar prize has been put on this problem.

The P = NP Problem

Suppose there are fifty cities to visit, and the distances between each pair of cities are given in a table. Since it is too difficult to find the optimal route—one in which every city is visited exactly once and the shortest possible distance is traversed—one could ask instead the following question:

> Is there a round trip covering every city that is at most 2,000 miles long?

There are two remarkable aspects to this problem:

- It is hopeless to try to solve this "easier" problem by examining all possible routes. For the first leg of the journey, one has 50 possibilities, then 49 for the second, then 48, and so on. Altogether there are

$$\begin{aligned} 50 \cdot 49 \cdots 2 \cdot 1 = {} & 30{,}414{,}093{,}201{,}713, 378{,}043{,}612{,}608, \\ & 166{,}064{,}768{,}844{,}377{,}641{,}568{,}960{,}512, \\ & 000{,}000{,}000{,}000 \end{aligned}$$

  possibilities, and trying to analyze all of them would overtax even the fastest computer.
- With luck, we might nevertheless be able to answer the question. Simply pick a route that looks like a relatively short one and check whether it is shorter than 2,000 miles. If it is, we have succeeded.

Put another way, we are facing a problem that one can solve only with a large amount of good luck. No one expects that a solution can be found quickly without luck. That is, no one believes that it is possible for anyone to think up a procedure that in a reasonable amount of time will solve the problem. Scandalously, however, as of yet, no one has been able to prove this assertion. Specialists speak of the "P = NP" problem, where the P indicates that a rapid procedure exists,[1] and NP means that a proposed solution can be shown to be correct or incorrect in a reasonable amount of time. There is a prize of one million dollars for anyone who can settle the "P = NP" question.[2]

An Example

In the following example, twenty "cities" were generated on a computer using a randomizing algorithm. Then a proposed route (see Figure 1) was generated using the method of *simulated annealing*, which is described in Chapter 60.

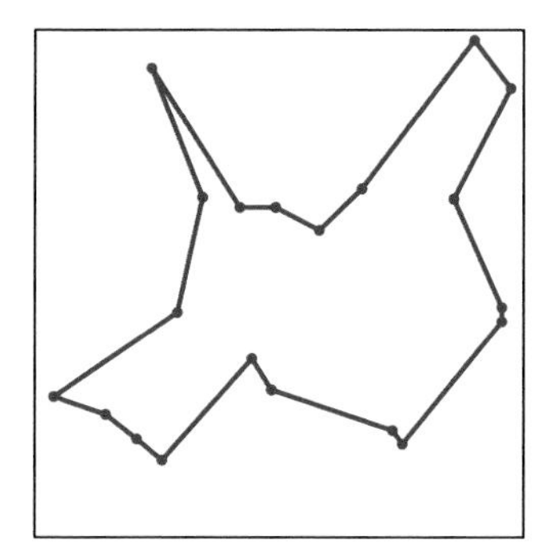

Figure 1. A proposed route for the traveling salesman.

[1]More precisely, one should be able to find the result in an amount of time that is a polynomial function of the size of the input.

[2]More on this problem can be found in Chapter 87.

Chapter 33

Squaring the Circle

The mathematical phrase "squaring the circle" has entered common usage to mean the conquering of an almost unsolvable problem. For mathematicians, these words evoke an exciting tale that fascinated both professionals and amateurs for over two thousand years.

Our story begins in Greece, where geometry was established on a firm foundation in Euclid's *Elements*. Ancient Greek mathematicians expended a great deal of energy in figuring out what lengths could be constructed from a given length using only a straightedge and compass. This restriction, which today seems rather arbitrary, was connected with the notion that the line and the circle were perfect forms.

Many of us learned how to carry out such constructions in high school: how to bisect an angle, draw a regular hexagon, erect right triangles on a given hypotenuse using Thales circles, and much more.

Several problems were posed by the Greeks that seemed to be of a more difficult nature altogether. One of them was to construct a square whose area was the same as that of a circle with given radius. This problem, the *squaring of the circle*, failed to yield to two thousand years of attempts. It was only in 1882 that the problem was finally settled, and surprisingly, it came about not through geometry, as most had expected, but via algebra.

Over the centuries, algebraists had analyzed the nature of numbers with great diligence and had determined a precise sense in which some numbers are "easy" and others "difficult."[1] It was long known that only certain "easy" numbers are constructible with straightedge and compass[2] and that squaring the circle could be shown to be unsolvable if one could show that the number π was, as generally suspected, a "difficult" number. Many mathematicians worked on this question, and in 1882, the mathematician Carl Louis Ferdinand von Lindemann published a proof that π is indeed a "difficult" number. His name will always be connected with this result.

In contrast to real life, in mathematics, "squaring the circle" is indeed impossible.

Construction with Straightedge and Compass

Here we shall look more closely at construction using straightedge and compass. The tools are a sheet of paper, a compass for drawing circles, and a straightedge, or unmarked ruler, for drawing straight lines. The sheet of paper comes with a line segment drawn on it of unit length. It is then an easy matter to construct a line of length 2: simply draw a line segment with the straightedge, and setting the compass to the length of the given unit segment, mark off two adjoining segments of length 1. Continuing in this manner allows you to construct the lengths $3, 4, 5, \ldots$. Likewise, one can easily construct the sum of any two lengths that have already been constructed, and, by going in the opposite direction, their difference.

Now ratios of sides in similar triangles come into play. We consider two line segments emanating from a point and cut by two parallel lines (see Figure 1).

Since the bigger triangle is similar to the smaller one, we have

$$\frac{x+y}{x} = \frac{b+a}{b},$$

[1] More will be said about these numbers, which are called *algebraic* and *transcendental*, respectively, in Chaper 48.

[2] To say that a number n is constructible means that given a line segment of arbitrary length 1 unit, a segment of length n can be constructed with straightedge and compass.

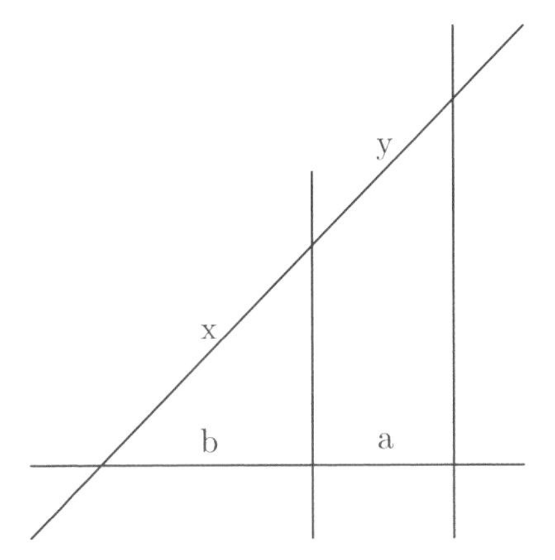

Figure 1. Division with the help of similar triangles.

from which we can conclude with a bit of algebraic manipulation that

$$\frac{x}{y} = \frac{b}{a}.$$

If we let y represent the unit length, and a and b lengths that have already been constructed, then x has length $\frac{b}{a}$. What this means is that for any pair of constructible lengths, their quotient is constructible as well. This holds as well for the product, since one could take a for the unit length and prescribe b and y. Then we would have $x = b \cdot y$.

Taking all the above considerations together, we conclude that we can construct all lengths that can be obtained from already constructible lengths using the operations of addition, subtraction, multiplication, and division.

But that's not all. There are more lengths that can be constructed: one can also extract certain roots. To see this, consider the right triangle pictured in Figure 2. It is well known that the square of the altitude is the product of the lengths of the two segments into which the altitude divides the hypotenuse: $h^2 = p \cdot q$.

Therefore, if p and q are lengths that have already been constructed, we can construct the right triangle with hypotenuse $p + q$ and altitude h of Figure 2 via the construction indicated in Figure 3.

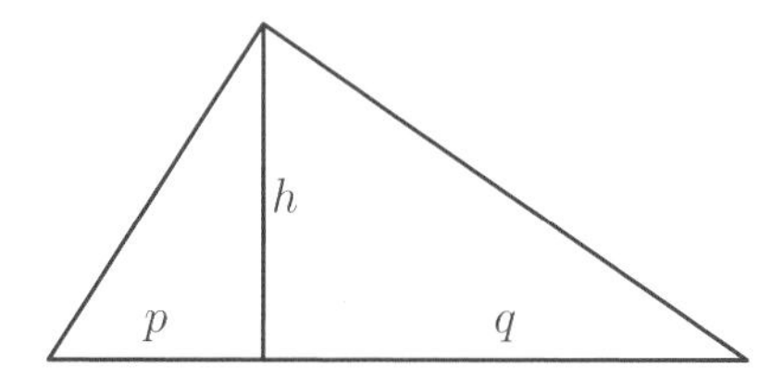

Figure 2. In this right triangle, $h^2 = p \cdot q$.

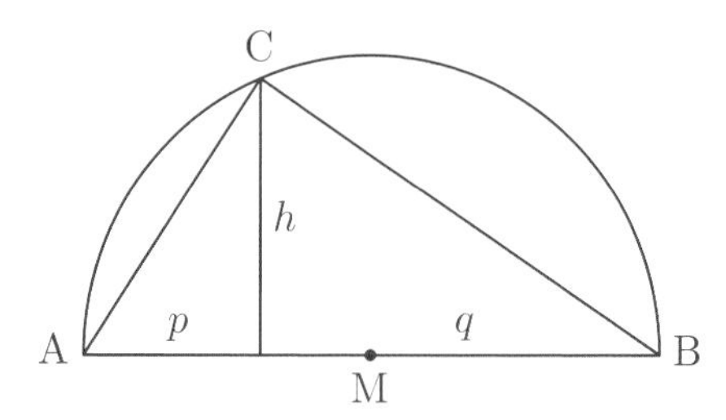

Figure 3. Extracting the square root using similar triangles.

The construction goes like this: Construct the line perpendicular to the segment AB as shown in Figure 3. Then construct M, the midpoint of AB and draw the semicircle with center M as shown. Label the point where the semicircle intersects the perpendicular C. From elementary geometry we know that triangle ABC is a right triangle. It follows that the two smaller triangles are similar to each other and to the big triangle. Therefore, using similar triangles, we obtain the relation $h/q = p/h$, which can be solved with simple algebra to yield $h^2 = p \cdot q$. And therefore h is the square root of $p \cdot q$. If we now let p be the unit length, then we have constructed $\sqrt{q}$.

Combining everything we have achieved thus far, we can construct rather complicated numbers, in fact, any number that can be formed from the number 1 and the operations $+$, $-$, $\cdot$, $\div$, $\sqrt{\ }$, for example, the number

$$\frac{\sqrt{3-\sqrt{2}}}{5} + 6.$$

Since the square root of the square root of a number is the fourth root, we can also without difficulty construct fourth, eighth, and sixteenth roots, and so on for every power of two. The procedures that we have sketched appear to allow us to construct numbers of arbitrary complexity. Why, then, shouldn't π be such a number, a number that eventually, perhaps on a very large sheet of paper, could be written as a combination of whole numbers with the symbols $+$, $-$, $\cdot$, $\div$, $\sqrt{\ }$? It is the result of Lindemann, whose proof goes far beyond the level of this discussion, that excludes such a possibility. All constructible numbers are much "simpler" than the number π.

Construction: With Straightedge and Compass Alone

The condition "with straightedge and compass alone" is to be read strictly: for example, if one allows markings on the straightedge, the situation changes dramatically. To illustrate the differences, we will show how a straightedge with markings on it can be used to divide an arbitrary angle into three equal parts. Under the strict rules of straightedge and compass *only*, such a construction has been proven, using algebraic methods, to be impossible.[3]

To see what is at issue here, let us consider an angle CBA (see Figure 4). As usual, this angle is denoted by three points, with the middle point as vertex.

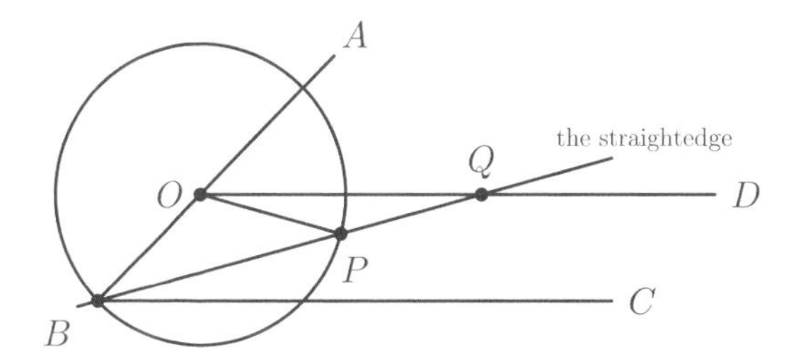

Figure 4. A neusis construction for trisecting an angle.

This angle is to be *trisected*—divided into three equal angles—and we assume that our straightedge contains two marks P and Q. We first measure off the length PQ on the ray BA, beginning at the

[3]Such constructions—straightedge and compass with a small amount of "cheating"—were investigated with great energy in the seventeenth century under the name *neusis* constructions.

point B. This gives us the point O, which we use as the center of a circle of radius PQ. Of course, the circle passes through B. Starting at O, we construct the parallel to BC, giving us the ray OD.

Now we use the marked straightedge. We place it so that it passes through B, the mark P touches the circle, and Q lies on the ray OD, as shown in Figure 4.

We are essentially done at this point. We maintain that the angle POQ is precisely one-third the angle CBA.

To prove this, it will be convenient to assign a name to the angle POQ. We shall call it α (alpha). We first observe that the angle PQO is equal to α. This is so because triangle OPQ is *isosceles*, the sides PO and PQ having the same length (equal to the radius of the circle), and therefore the angles opposite those sides must be equal.

Since the sum of all the angles in triangle OPQ is $180°$ and we know two of the angles, it follows that the angle OPB must equal 2α, since the angles OPB and OPQ add up to $180°$.

Triangle BPO is also isosceles: OP and OB are equal to the radius of the circle. We conclude that the angle OBP also must have the value 2α.

The last step of the proof is to observe that the angle QBC is equal to α. This angle must agree with the angle OQB, since they are alternate interior angles formed by a line (the straightedge) cutting two parallel lines (the rays BC and OD). Therefore, the angle OBC, the sum of the angles QBC and OBQ, has measure 3α, which means that the angle POQ is one-third of the angle CBO.

Cubing the Sphere?

As we have seen, squaring the circle is impossible under the rules that restrict one to straightedge and compass, and as we mentioned above, in common parlance, the difficulty of a task is often compared with that of squaring the circle.

During the protracted negotiations at the end of 2005 to form a new German governing coalition, the chancellor-designate Angela Merkel, wishing to make a point, said to the press that the negotiations were more difficult even than squaring the circle, perhaps as difficult as cubing the sphere. We may assume that she was referring to the problem of transforming a sphere into a cube of the same volume. Since a sphere of radius r has volume $\frac{4}{3} \cdot \pi \cdot r^3$ and a cube of side length ℓ has volume ℓ^3, one must have

$$\frac{4}{3} \cdot \pi \cdot r^3 = \ell^3,$$

or equivalently,

$$\ell = \sqrt[3]{\frac{4}{3} \cdot \pi} \cdot r.$$

In other words, cubing the sphere requires the construction of $\sqrt[3]{\frac{4}{3}\pi}$. If that were possible, then with the methods we have described, one could construct π, and then the squaring of the circle would be no problem.

The converse is false, however, since in general, one cannot construct cube roots with straightedge and compass.

Therefore, Angela Merkel was correct in implying that the problem of cubing the sphere is more difficult than that of squaring the circle (although one could argue contrariwise that it makes no sense to talk about comparative degrees of difficulty when both constructions are impossible).

Chapter 34

A Step into the Infinite

How can one grasp the infinite? How does one prove, for example, that for *any* positive integer n, the sum of the first n integers is always equal to $\frac{1}{2} \cdot n \cdot (n+1)$? Let us begin by testing whether the assertion has even a chance of being true by checking a few examples. If $n = 4$, we have that the sum of the first n positive integers is $1+2+3+4 = 10$, and indeed, substituting $n = 4$ into the formula $\frac{1}{2} \cdot n \cdot (n+1)$ yields $\frac{1}{2} \cdot 4 \cdot (4+1) = 10$. We could test other numbers as well, but no matter how many we test, how can we be sure that the formula is *always* correct? Even for ten-thousand-digit numbers, even for those that to write down would take all the ink manufactured in a year?

The answer is certainly not to begin testing all possible numbers. Even if you could network all the world's computers for the calculations, you wouldn't even get as far as 20-digit numbers.

So how, then? Mathematicians prove the correctness of this and similar assertions using a technique called *induction*. A proof by induction requires two things: First, one must do a calculation and prove the assertion for the smallest value of n under consideration, which in our case is $n = 1$. But that is easy, since the "sum" of the first one number is just 1, and putting 1 into the formula yields $\frac{1}{2} \cdot 1 \cdot (1+1) = 1$. Second, one must prove that *if* the assertion has been proved for some number, then it must also be true for the following number. (The calculation for the current case will be carried out below.)

Since the assertion holds for $n = 1$, then by the second step of the proof method it must be true for $n = 2$ as well. And if it is true for $n = 2$, it must be true for $n = 3$. But if it is true for $n = 3$, it must be true for.... As an analogy, consider a row of dominoes standing in such a way that if one is tipped over, it will tip over its neighbor. Then if the first domino is tipped, all the dominoes will eventually fall over.

What is of greatest interest about induction is that in a single proof occupying a few lines of print one can establish the truth of an infinite number of statements. It is the key to almost all mathematical assertions that need to be proved for an infinite number of cases.

The Missing Induction Step

Here we shall carry out the so-called induction step for the proof of the summation formula given above.

We need to show that the sum of the first $n+1$ numbers, that is, $1+2+\cdots+n+(n+1)$ is given by the above formula if we are given the *induction hypothesis* that the sum of the first n numbers is given by the formula $\frac{1}{2} \cdot n \cdot (n+1)$.

Given the induction hypothesis, we conclude that $1+2+\cdots+n+(n+1) = \frac{1}{2} \cdot n \cdot (n+1) + (n+1)$. But this expression is equal to $\frac{1}{2} \cdot (n+1)(n+2)$ (this is easily checked by algebraic manipulation), and that is the summation formula for the case $n+1$.

We have shown, then, that if one assumes the formula for n, it must also be true for $n+1$.

Where Does the Formula Come From?

Induction is the "official" way of validating assertions about an infinite number of natural numbers. But before one can begin to prove an assertion, one must have an assertion to prove! How does one discover such things?

This is the creative aspect of mathematics. One needs intuition, experience, luck, and frequently a clever way of visualizing the problem. Let us see how this works with our standard example, the assertion about the sum of the first n natural numbers:

$$1 + 2 + \cdots + n = \frac{n \cdot (n+1)}{2}.$$

We have seen the proof, so now let us see where the formula came from. There are many ways of arriving at the formula, and we will look at two of them.

One possibility is to imagine the problem as a sum of areas, as shown in Figure 1.

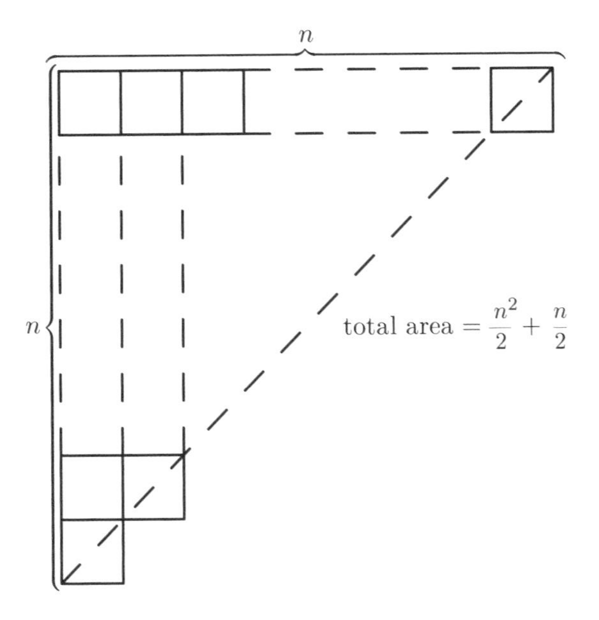

Figure 1. One way of visualizing the formula $1 + \cdots + n = n(n+1)/2$.

We begin with the small square in the lower left-hand corner of the figure. We place two squares on top of it, as shown, then three, all the way up to n squares in the top row. The result is something like half a chessboard. Now, half a chessboard has area $n \cdot n/2$, but the area of our squares exceeds that of the half-chessboard by half a square for each square on the diagonal. We therefore need to add an area of $\frac{1}{2}n$, and the result is that $1 + \cdots + n$ should equal $n \cdot n/2 + (1/2)n$. And this expression is equal to $n \cdot (n+1)/2$.

Alternatively, we could proceed as the schoolboy Gauss did in the anecdote presented in Chapter 25: We write the sum $1 + \cdots + n$ as $(1+n) + (2+(n-1)) + \cdots$ by collecting the first and last summands, the second and second-to-last, and so on. Each term in the (larger) parentheses is equal to $n+1$, and the number of terms is $n/2$ if n is even. If n is odd, there are $(n-1)/2$ such terms together with the lone term $(n+1)/2$ in the middle.

Altogether, for n even, the sum should be $(n+1) \cdot n/2$, while for n odd, it should be $(n+1) \cdot (n-1)/2 + (n+1)/2$, which after a bit of algebraic manipulation also takes the form $(n+1) \cdot n/2$. Thus we obtain the same formula for the sum of the first n natural numbers regardless of whether n is odd or even.

An Additional Proof by Induction

As a further example of a fact that can be proved by induction, let us consider the statement, "n objects can be arranged in $1 \cdot 2 \cdot 3 \cdots n$ different ways." To put it more precisely, this means that for $n = 1$ there is one arrangement; for $n = 2$ there are $1 \cdot 2 = 2$; and so on. We observe as well that $n!$ is the standard abbreviation for $1 \cdot 2 \cdot 3 \cdots n$ and call this number "n factorial."[1]

For small n this can be checked directly. For example, for three objects a, b, c, the possible arrangements are abc, acb, bac, bca, cab, cba. And indeed, there are $3! = 6$ of them.

A "rigorous" proof by induction might look like this: *First* one shows that the assertion is true for $n = 1$. This is clear, since with a single element, there is only one possible arrangement. *Second*, a value of n—which can be thought of as an arbitrary positive integer, no matter how large—is fixed, and it is assumed (the induction hypothesis) that the statement is true for that number. *Third*, one shows that given the induction hypothesis, the assertion is true for $n+1$ objects.

[1] See Chapter 29 for more on arrangements.

Let us now proceed. We imagine the n objects as a collection of white marbles numbered 1 through n. The $(n+1)$th object will be a red marble. In ordering *all* $n+1$ marbles, we proceed as follows. We first choose an ordering for the n white marbles, of which there are $1 \cdot 2 \cdot 3 \cdots n = n!$ by the induction hypothesis. We then consider what to do with the red marble. It could be placed at the beginning, after the first marble, after the second, after the third, and so on, with the final option to place it at the end of the row. That makes $n+1$ possibilities for the red marble. Since the red marble offers $n+1$ possibilities for *each* possible arrangement of the n white marbles, there are altogether $(n+1) \cdot n!$ arrangements of the $n+1$ marbles, or $(n+1)!$, which is precisely the assertion for $n+1$ objects.

Chapter 35

Mathematics in Your CD Player

Among all the technical gadgets that are found in the average home, the CD player is certainly the one that contains the most mathematics. It is important in two respects. The first is that the original continuous signal—the playing of the Berlin Philharmonic, for example—is digitized and thereby converted into a finite collection of zeros and ones. The signal is then sampled about 44,000 times per second, and an important theorem of signal processing ensures that at that level of sampling, everything that the human ear is capable of hearing has been captured. (If our hearing were significantly better or worse, in terms of the frequencies that we can hear, the CD player would have different parameters.)

A further requirement of a CD player for mathematics results from the fact that neither the process of pressing a CD nor that of playing it back is free from error: perhaps a speck of dust landed on the disc, or the cat scratched it. This is a big problem, as anyone can imagine who has experienced the loss of an entire computer file as a result of a single bit of information among millions being incorrectly transmitted.

If one hoped to achieve comparable perfection with a CD player, both apparatus and discs would be unaffordable. But the problem was solved by different means, and the magic word here is *coding theory.*[1] How can one transmit a digital message in such a way that the recipient is able to read it even if there have been errors in transmission?

How would you, for example, send a ten-letter text in Morse code so that even if it becomes distorted due to miscoding or an atmospheric disturbance, the original ten letters will be guaranteed to arrive intact? One idea that may spring to mind is to send the message repeatedly, and the recipient should choose the version most frequently received as the correct one. That would be much too much work for the CD player, and therefore, techniques have been developed so that a "robust" version of the actual transmitted signal is not significantly longer than the original.

In the meantime, the modes of transmission have become so impervious to error that the quality of sound reproduction doesn't suffer even in the face of considerable disturbance. For instance, even a badly scratched CD can be listened to with no detectable imperfection. It is a pity that such techniques were unavailable for vinyl records; in the old days, one could hear every speck of dust.

The Sampling Theorem

For music or some other acoustic signal to find its way from the original source to your home stereo system, the following steps are taken. First, the sound is *digitized*, that is, converted into a very long stream of zeros and ones. This transformation from the analog, or continuous, world to the digital, or discrete, is a decisive step, for it is only after digitization has occurred that the material can be copied and recopied and processed without any loss of quality.

[1]More on coding theory can be found in Chapter 98.

The success of digitization is possible only because we human beings do not hear perfectly. In a world in which we could hear arbitrarily high frequencies, there would be no CDs. But in fact, frequencies above about 20 kilohertz are inaudible to us, and so successful digitization can take place. It proceeds in two steps:

- First the signal is sent through a filter, in which all frequencies above some inaudible (to us) frequency are suppressed. We would be unable to hear the difference between the two signals.
- Then one makes use of the fact that a frequency-limited signal can be digitized and then accurately reconstructed if it is sampled enough times per second.

The fact mentioned in the second point above is known as the *sampling theorem*. Here is a precise formulation:

> If a signal is composed of a number of frequencies, the highest of which is f, then the digitized signal can be reproduced if the time between samples of the signal is at most $1/(2f)$.
>
> For example, if the highest frequency to occur is 10 kilohertz, then a sampling distance of 1/20,000 is necessary, that is, a rate of 20,000 samples per second.

While this may seem rather abstract, the theorem can be illustrated in another realm altogether. Suppose you have a camcorder that allows you to set the number of frames per second. Your toddler is sitting on a swing, and you would like to film her. At a normal frequency, your child's swinging will be realistically reproduced. But if you set the frequency too low, enormous distortion can result: between one picture and the next the swing may appear to have moved only a small amount, while in reality it has made a complete cycle that the camcorder missed. The sampling theorem can be seen as nothing more than a sort of user's manual: you should select such-and-such frequency so that swinging at such-and-such a rate will be accurately reproduced.

Chapter 36

The Logarithm: A Dying Breed

Older readers will recall from their schooldays—perhaps with horror—how they learned to calculate with logarithms. Here we present an obituary to the logarithm. To be sure, logarithms remain an important part of mathematics, but in the world of engineers and technicians, they are threatened with extinction.

To understand their utility, one must recall some terminology. First, one must know how mathematicians notate powers: if a and n are integers, then a^n denotes the n-fold product of a. Thus 3^4 represents the product $3 \cdot 3 \cdot 3 \cdot 3$ (which equals 81), and 10^6 is equal to one million. If one multiplies a number by itself n times and then an additional m times, one has $m + n$ factors altogether, and this observation yields the power law $a^{n+m} = a^n \cdot a^m$. The same power law holds even after one defines raising to a power (exponentiation) for an arbitrary exponent. For example, the square root of a can be written $a^{1/2}$.

And now logarithms make their appearance. For the sake of simplicity, we shall consider only logarithms to base 10. For a number b, the logarithm of b is defined as the number m with the property that $10^m = b$. We have seen above that the logarithm of one million is 6, and the logarithm of one thousand is certainly 3. What is important is that every positive number has a logarithm when one allows arbitrary exponents.

The crucial point regarding logarithms is that by the power law derived above, the logarithm of a product is equal to the sum of the logarithms of the factors, and this fact allows multiplication to be transformed into addition. If the product $b \cdot c$ is to calculated, one can determine the logarithms of b and c with the help of a table of logarithms, add the two numbers, and then look up what logarithm belongs to this sum. That is the product.

Once upon a time, before there was a computer on every desk, that is how multiplication was frequently done; logarithms were the workhorse of calculation. Because adding is much easier than multiplication, the transformation from one type of calculation to the other made life easier for those who had to do arithmetic with paper and pencil. Such a technique has analogues in other aspects of our lives, where a problem might have to be translated into a different idiom before it can be solved.

And now for another obituary, this time one that is final and irrevocable: Logarithmic calculation was mechanized in a most user-friendly way in the form of the slide rule (see Figure 1). These are truly extinct, to be found only in museums of technology.

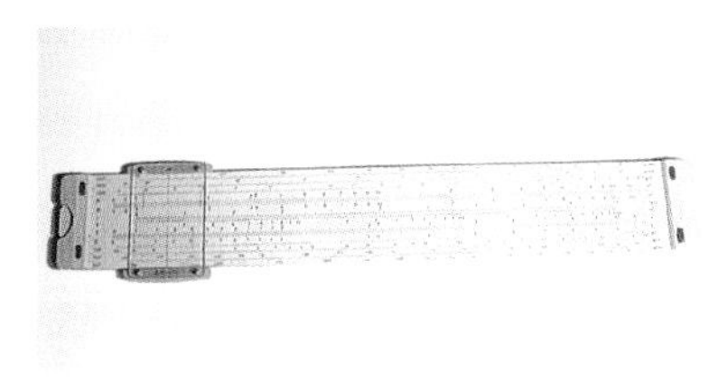

Figure 1. A slide rule.

A Typical Calculation

Let us follow our obituary with a sample calculation, just as it was done in the time before everyone had a pocket calculator within reach.

Suppose we wish to calculate $3.45 \cdot 7.61$. Since we don't yet know the answer, let us call it x:

$$x = 3.45 \cdot 7.61.$$

By the power law, the logarithm of x is equal to the sum of the logarithms of 3.45 and 7.61. Let us use base-10 logarithms, as some of us did in our schooldays. Then the logarithm of 3.45 is 0.53782, and the logarithm of 7.61 is 0.88138. These numbers would have been looked up in a table of logarithms. It follows that the logarithm of the (as yet unknown) number x is equal to

$$0.53782 + 0.88138 = 1.41920.$$

Since from the definition of the logarithm we have $10^{\log x} = x$, it follows that

$$x = 10^{1.41920} = 26.25427,$$

this last number having been found as well in a table of logarithms, where one uses reverse lookup: what number x has logarithm 1.41920?

The result is quite good, since the exact value is

$$3.45 \cdot 7.61 = 26.2545.$$

Therefore one can forget about multiplying and do everything with addition once one has translated the problem into the world of logarithms.[1]

[1]Note: Since logarithms are dying from popular consciousness, your translator cannot resist one last opportunity to recount a joke that was told to him about forty-five years ago, when he was a lad, by one of his mathematics teachers. It goes like this: After the flood waters had subsided and the ark come to rest on Mount Ararat, Noah sent the animals out into the world two by two with the injunction to be fruitful and multiply. All the animals left the ark, and eagerly began to obey Noah's command, except for two small snakes, who shyly approached Noah and said, "What about us? We cannot multiply." "And why is that?" asked Noah. "Because we are adders," they replied. At first, Noah was nonplused, but then he had an idea. He went into the forest, cut down a few saplings, and lashed them together to build a table. Placing the snakes on the table, he encouraged them with these words: "Now be fruitful and multiply, for with a log table, even adders can multiply."

Chapter 37

Prizeworthy Mathematics

What? You haven't heard about prizes for mathematicians? That is probably because as a rule, there is not much money at stake, and even if there were, it is usually so difficult to explain mathematical accomplishments to the general public that the news media don't bother reporting the story.

Figure 1. The most prestigious mathematics prize: the Fields Medal.

Here is a brief summary of the most important prizes for mathematical achievement. Any mathematician who wishes to achieve immortality should accomplish something spectacular while still relatively young. Mid-twenties is about right. Then there is the possibility of a Fields Medal, which is awarded every four years at the

International Congress of Mathematicians.[1] Such a prize will not bring great wealth—the prize money is about twenty thousand dollars. Nevertheless, the winners can be sure of financial security for the rest of their lives, since they will be certain to obtain the best positions and be barraged with offers for lucrative speaking engagements. The Fields Medal is generally viewed as a sort of "Nobel Prize in mathematics." However, the prize is strictly limited to mathematicians who have not attained their fortieth birthday. (Thus there was no Fields Medal awarded at the 1998 congress in Berlin for perhaps the most exciting mathematical discovery of the century, the proof of Fermat's last theorem, by the forty-five-year-old Andrew Wiles.)

There do, however, exist prizes that carry a hefty monetary award. For example, there are the million-dollar prizes offered since 2000 by the Clay Mathematics Institute for solutions to seven concrete difficult problems.[2] All are still open, despite the fact that many of the world's best mathematicians have been grappling with them.[3]

And since 2003, a prize has been offered with an award set at the level of the Nobel Prizes. The Abel Prize was financed by a wealthy Norwegian, and perhaps someday it will be drawn into the orbit of the Nobel Prize and be awarded at the annual ceremony in Sweden. The first recipient of the award was the French mathematician Jean-Pierre Serre, and it was not an easy task for the prize committee to try to explain to a lay audience what Serre had done to merit an award of $800,000. Perhaps that is because all the sciences have become highly specialized: do you remember what last year's Nobel Prize in chemistry was awarded for?

---

[1] Translator's note: The Finnish mathematician Olli Lehto has written a fascinating history of the International Mathematical Union, which sponsors this congress: *Mathematics without Borders*, Springer, 1998.

[2] Translator's note: These problems are expounded in Keith J. Devlin's *The Millennium Problems*, Basic Books, 2002.

[3] Translator's note: This is no longer the case! See Chapter 93, on the Poincaré problem.

Getting Rich through Mathematics: Does an Amateur Have a Chance?

There are many mathematical problems that some of the world's best brains have been struggling over for decades. Some of them have been described in this book (see Chapters 18, 32, 49, 57).

For some of them, a fat purse is being offered, and so in addition to undying fame, one might take home a million dollars. Does a mathematically inclined amateur have a chance? The history of mathematics has a number of episodes in which respectable, even outstanding, results were achieved by mathematical hobbyists. In this book, Bayes and Buffon belong to this group (Chapters 50 and 59). Technically, also the great Pierre de Fermat, a jurist, was not a professional mathematician.

However it is highly unlikely that the truly difficult problems will be solved by an amateur. The level is simply too high, and all obvious approaches have almost surely been tried.

In other areas of human endeavor as well, the greatest achievements are not to be expected from those not active in the field at the highest level. The Wimbledon tournament will surely not be won by someone who plays tennis occasionally on weekends. And the role of Siegfried will certainly not be sung at Bayreuth by a tenor who does not have years of study behind him.

Chapter 38

Why Axioms of All Things?

Children aged about three to six often get on their parents' nerves with their endless barrage of innocent questions, such as "Mommy, how does a car work?" but beyond the terms "motor," "combustion," "chemical reaction," even experts quickly get to the point of, "That's just the way it is!" End of discussion.

Just so with mathematics. One can keep asking more and more basic questions, but since at some point the discussion becomes fruitless, agreement has been reached on a starting point, where things are "just the way it is." These are the axioms of mathematics.

The first axiomatic system was propounded over 2,000 years ago by Euclid (third century B.C.E.; see Figure 1). In his *Elements*, he gave an axiomatic basis for geometry. First of all, there are fundamental concepts such as *point* and *line*, and then certain properties are defined for these objects: "Through every pair of points there is exactly one line." And so on.

Today, almost all areas of mathematics have been axiomatized. There are axioms about all sorts of objects, such as numbers, vectors, and probabilities.

Figure 1. Euclid in the "School of Athens" (Raphael, 1510).

After the axioms have been established, one has free rein to explore the possible consequences. Most mathematicians find this much more exciting than endlessly quarreling over the foundations.

But one thing remains a mystery. How does it happen—for example in geometry—that with a few axioms one can develop a theory that gives a superbly functioning mathematical model for a description of the phenomenal world? The great success of the axiomatic method has resulted in thinkers in other fields of human endeavor attempting to establish an axiomatic basis for their science. It is thus that Newton's great work on mechanics (significantly titled *Philosophiae Naturalis Principia Mathematica*) reads over long stretches like a mathematics textbook.

If one replaces the word "axiom" by the term "rules of the game," one obtains a quite useful analogy. The rules of a game—chess, for example—are fixed, and the players' intellectual energy is not spent in trying to figure out new and improved rules. Rather, their efforts are in figuring out whether, for example, in a certain position a checkmate is possible. Just so, mathematicians wish to know whether under a particular set of axioms, a certain problem has a solution.

Hilbert's Program

Although the first axiomatic systems were created over two thousand years ago, the triumphal progress of this approach began only about a century ago. It was the great mathematician David Hilbert who suggested that mathematics should be placed on a solid axiomatic footing. The basic idea was that this would achieve two goals. First, it should then be possible to derive mathematical theorems mechanically from these axioms using the rules of deduction, and second, one should be be able to decide, again mechanically, on the truth or falsity of a given mathematical statement. For example, "yes" would be the result of asking the question, "Is there an integer whose square is 25?" while "no" would follow from "Is there a solution to the equation $x = x + 1$?"

This ambitious program ended in failure. The logician Kurt Gödel proved in his incompleteness theorems that in a mathematical theory, one can never be certain whether some statement and its negation might both be derivable from the axioms. These theorems proved as well that there will always be true statements that cannot be derived from the axioms.

Axioms Are the "Laws" of Mathematics

In addition to the analogy given above in which the rules of chess were likened to the axioms of mathematics, one might also find parallels in the legal field. Once laws have been established, there are certain things that one may do without fear of punishment. For the lawyer, the moral judgment of such actions is secondary to the question whether a person's actions are legally sanctioned. Similarly, in mathematics, it is difficult to decide how a subject will develop from the given axioms. Perhaps undesirable behavior results from a certain set of laws, in which case the laws can be rewritten or modified. Just so in mathematics, one can modify the system of axioms.

In law, in contrast to chess, the moral component is significant, and such is the case for mathematics as well, since a mathematician's discoveries can be used for the good of mankind as well as for nefarious purposes. An optimization theorem, for example, can be used to create improved fertilizers, but also to produce biological weapons.

Chapter 39

Proof by Computer?

Today we are going to discuss a problem that borders on the philosophical, one that mathematicians are having to confront as technology develops ever more rapidly: under what circumstances can a mathematical assertion be said to have been definitively proved?

Over the past two thousand years, consensus has been reached as to what constitutes a "rigorous proof" in a science whose fundamentals have been established: the assertion to be proved must be derived by correct deductive reasoning from the foundational axioms of the science. That is how Euclidean geometry was established, and many other fields developed along the lines laid down in Euclid's *Elements*. It was a long time before all the areas of mathematics were established on a firm axiomatic footing, but by the middle of the nineteenth century, the foundations of mathematics were fairly well settled. What was considered correct, what assertions had been proved, depended on the consensus of the mathematical community, on what had received the blessing of the experts.

In the 1970s, these standards were called into question by the proof of the renowned "four color problem."[1] For the first time in the history of mathematics, computers played a crucial role, since

[1] The problem involves the coloring of maps; its details are irrelevant to the present discussion. More on the four color problem can be found in Chapter 99.

significant portions of the calculations, performed by computer, could not possibly be carried out "by hand" in a human lifetime.

Was it then a valid proof? The mathematical community is divided. Most seek to avoid computer proofs, and when a theorem has been proved with the aid of a computer, intensive effort is devoted to finding a "classical" proof. Sometimes such a search has borne fruit, but in many cases one has had to rely on the flow of electrons.

This problem has yet another facet, since now computers can be programmed to find and prove their own, albeit simple, theorems. One must wonder whether this aspect of computer proof will develop along the lines of computer chess, which at the outset was rather primitive, easily beaten by even a club player, while today even world champions are checkmated. If that happens, mathematicians will have a problem on their hands.

It Is Obvious That...

The answer to the question, "Is this a valid proof?" depends not only on when it is asked, but also on the background of those who are discussing the matter. For example, if it is a problem concerning the natural numbers (that is, the numbers $1, 2, 3 \ldots$), then proofs are generally carried out by induction.[2] It is only with this method of proof that facts about the infinite can be ascertained with finite methods.

Those who in their mathematical lives have carried out this type of proof frequently prefer not to spend time and energy carrying out such a standard argument. Thus one frequently finds proofs of the form, "By induction it follows that...," or one even encounters a result that is simply used without comment.

Such situations cause consternation among mathematical neophytes. At the beginning of their studies, they expect complete proofs, but the experts cannot be bothered with filling in every detail. Over time, the mathematician becomes used to a lack of exhaustive proof. While full details are to be preferred in doubtful situations,

[2] Compare Chapter 34.

it is questionable whether technicalities that readers have long since mastered should occupy a large portion of a scientific article.

Can Computers Be Mathematically Creative?

Before one can write a proof, one must first know just what it is one wishes to prove. For example, if one has no idea that an angle inscribed in a semicircle (such as angle C in Figure 3 of Chapter 33) measures 90°, then one would not even think of providing a proof for the fact.[3]

It is a consensus among mathematicians that this creative aspect of mathematical progress—where do statements come from that are to be proved?—is in many cases more demanding than the requisite proof.

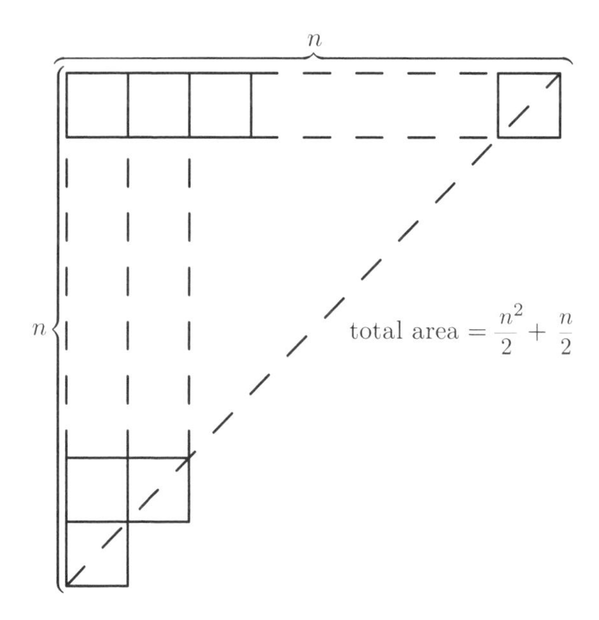

As an example, let us consider the statement that the sum $1 + 2 + \cdots + n$ of the first n natural numbers is equal to $n \cdot (n + 1)/2$.[4] It is decades since computers have been able to carry out in full rigor

[3]This is Thales' theorem, which is discussed in Chapter 47.
[4]More on this can be found in Chapter 34.

the proof by induction of this fact. But it is doubtful whether a computer could have discovered the fact itself. We human beings can picture the half-chessboard in the accompanying figure that reveals the correct formula. Computers, on the other hand, can "see" only what has been programmed into them, and therefore mathematicians are convinced that computers will never take over this interesting part of their work.

Chapter 40

The Lottery: The Small Prizes

When mathematicians are asked about the odds of winning the lottery, what the questioner usually wants to know is the probability of a ticket winning the jackpot. We have written about this in Chapter 1, where we saw that the odds of winning are a pitiful 1/13,983,816. A person buying one ticket per week would expect on average to have to wait about 270,000 years before selecting a winning ticket. A person wishing to have an expectation of one win by betting on the lottery every week for seventy years should plan on purchasing about 4,000 tickets a week.

Most of us have to be content with a smaller gambling budget, and we may console ourselves with the better odds of winning one of the smaller prizes. A lottery bet consists in choosing six numbers in the range 1 to 49. The jackpot is awarded to a ticket with all six correct, but there are prizes for three, four, and five correct. The odds of choosing three correct numbers are not tiny, about 0.018, or 1.8%. While a run of bad luck could mean not a single "three correct" in a whole year, one may console oneself with the fact that the probability of at least one "three correct" ticket out of 52 tickets is about 61%.

Getting four correct is of course more difficult to achieve. Only about one ticket in one thousand is such a winner. Five correct occurs in only about two tickets out of every hundred thousand. Of course, there is still the bonus number to consider, which is an additional number selected from the 43 numbers that were not selected as winning numbers. From a naive point of view, the bonus number should increase one's odds tremendously, but the mathematical truth is alas more sobering. One is only six times as likely to have five correct plus the bonus number than to have guessed all six numbers correctly. The difference is worse between the odds for four correct and three correct plus the bonus number. The probability increases by a factor of only 1.33.

However, as always, what you are buying with your lottery ticket is primarily a couple of days' worth of dreams about great wealth, and one may keep in mind that the money that is not paid out in prizes is spent on worthy social projects.

The Small Prizes: Calculations. The concepts developed in Chapter 29 allow us to compute more than just the probability of winning the grand prize. Recall that there are $\binom{n}{k}$ (read "n choose k") ways of selecting k members of an n-element set, and therefore there are $\binom{49}{6} = 13{,}983{,}816$ different lottery wagers.

Now we would like to calculate the probability of choosing exactly three out of six correct lottery numbers. How many ways are there, then, to choose six numbers out of forty-nine in such a way that exactly three are correct and three incorrect? What we have to do is to choose three numbers from the set of six "lucky" numbers, and three from the set of forty-three remaining "unlucky" numbers. Thus there are $\binom{6}{3} = 20$ ways of choosing three correct numbers and $\binom{43}{3} = 12{,}341$ ways of choosing three incorrect. Therefore, there are

$$\binom{6}{3} \cdot \binom{43}{3} = 20 \cdot 12{,}341 = 246{,}820$$

different ways of placing a lottery bet with three correct and three incorrect. Since there are 13,983,816 possible bets, the probability of three correct is

$$\frac{246{,}820}{13{,}983{,}816} = 0.0176466\ldots,$$

or just under 1.8%.

What is the probability of choosing no correct numbers? To do that, all six numbers must be chosen from the set of forty-three incorrect numbers. Therefore, we obtain a probability of

$$\frac{\binom{43}{6}}{13{,}983{,}816} = \frac{6{,}096{,}454}{13{,}983{,}816} = 0.43587\ldots.$$

Here is a complete table of probabilities for one through six correct:

k	Probability of k Correct
0	0.436
1	0.413
2	0.132
3	0.018
4	0.001
5	$2 \cdot 10^{-5}$
6	$7 \cdot 10^{-8}$

The Bonus Number

How do the odds improve with the bonus number? Let us count:

> A bet with five correct numbers can be chosen in $\binom{6}{5} \cdot \binom{43}{1} = 6 \cdot 43 = 258$ ways. Therefore, a five-correct ticket is 258 times as likely than a grand prize winner. A wager consisting of five correct plus the bonus number arises from a selection of five among the six correct numbers (of which there are $\binom{6}{5}$ possibilities) and simultaneously the unique bonus number (there is only one way to choose it). There are thus $\binom{6}{5} = 6$ possibilities.

Since there is only one way of choosing all six correct, one may say that one is six times as likely to choose five correct plus the bonus number as to guess all six correct.

Chapter 41

Formulas = Concentrated Thought

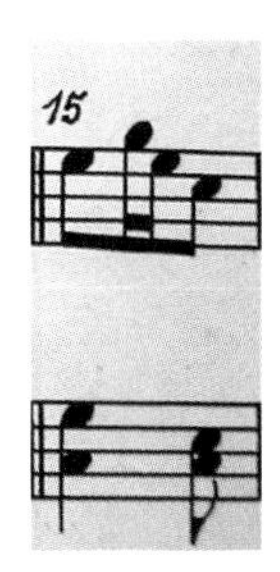

Formulas are the language of mathematics. Over the centuries, specialized notations have developed by which those in the know can communicate their thoughts with an economy of writing. Just as the notes of Beethoven's *Ninth Symphony* can be translated by an orchestra in Seoul into music as conceived by the composer, mathematical formulas are understood across cultural boundaries.

Like musical notation, mathematical formulas are an invention of modern times. Today's students of mathematics would have difficulty extracting the mathematical content from a sixteenth-century work by Adam Ries. While he wrote about algebra, he did not use formulas. Instead, all his calculations are written out "in prose," making very difficult reading for us in the twenty-first century.

> The statement that the equation $3 \cdot x + 5 = 26$ holds for some number x would have been expressed in one of Ries's books as follows: "Take a number and multiply it by 3. Then add 5 and you obtain 26. How large is the number?"
> The method of solution given is itself of interest: Ries used a technique whereby two different values for x are

> guessed and substituted into the problem; that is, for each choice, $3 \cdot x + 5$ is to be calculated. Depending on whether the results are too large or too small and by how much, these guessed values are interpolated to yield the exact value of x.

Using a specially adapted form of writing to communicate ideas is not unique to music and mathematics. One may be familiar, for example, with the notation for dance steps or for chess positions, blueprints, or chemical formulas. All those who use such specialized "languages" will agree that they make communication much easier. Even in creative thought, it is helpful if one can emphasize what is most important with a suitable notational system.

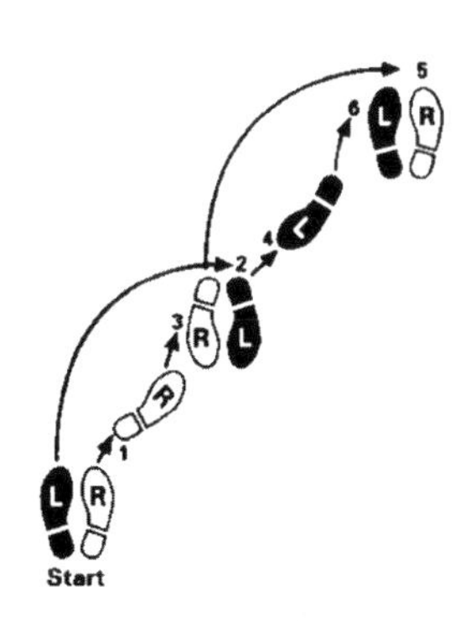

Of course, it depends on the reader's mathematical experience how quickly the content of a formula will be grasped, and the same holds for musical notation, chess positions, and so on.

Finally, it should be emphasized that it has never occurred to any mathematician that formulas are the essential content of mathematics. They are simply aids in the recording and transmission of ideas for oneself and one's colleagues. Similarly, no musician would maintain that the essence of understanding music consists in the ability to read and write notes.

Algebra Emancipates Itself from Geometry

It was a long journey in the history of mathematics to today's standard notation. Anyone in the Middle Ages who wished to express that one was looking for a number x such that $x^3 = 5$, for example because one wished to know the side length of a cube of volume 5, could do so only in prose: "What value does a number have that multiplied by itself three times gives the number 5?" One may imagine the difficulty in expressing complex computations, such as the calculation of interest.

The first important steps toward a better system of notation come from the Italian Renaissance. By the eighteenth century, a standard had been achieved that is essentially what is in use today. At that time, the most important mathematical constants were named: In 1731, Euler used the symbol e to denote the base of exponential growth.[1] From England came the idea of using π to denote the ratio of a circle's circumference to its diameter, apparently because the Greek pi corresponds to the Latin letter P, which can be taken to stand for "perimeter."[2]

There was an additional problem that hampered the development of mathematical notation. Until the modern era, calculations were almost exclusively geometrically oriented. Expressions such as x^5 and $x^3 + x$ were considered nonsensical, the former because it implied the existence of a five-dimensional object, and the latter because it appeared to be attempting to combine a three-dimensional volume (x^3) with a one-dimensional length (x), like trying to add apples and oranges.

It was Descartes who freed mathematics from its enslavement to a geometrical interpretation. Suddenly, mathematical theorems had a whole new range of application. Today, problems are formulated and solved involving thousands of variables, for example in the optimization of a subway schedule. And to think that only a few centuries ago, x^5 caused some of the cleverest heads to be scratched in wonder.

---

[1] See Chapter 42.

[2] "Circumference" is the special name given to a circle's perimeter.

Chapter 42

Endless Growth

These are tough times for investors as interest rates continue at record lows. Let us imagine a bank in some fictitious land that offers the fabulous rate of 100%: after one year, a one-euro deposit becomes two euros. Our friend Isabella has the bright idea of taking full advantage of the situation.[1] She withdraws her deposit after half a year, now worth 1.5 euros, and immediately redeposits it. After another half year, she withdraws again. In this second half-year, her deposit again increases by a factor of 1.5, giving her a total of 2.25 euros. If she were to increase her frequency of withdrawal and redeposit to four times a year, Isabella would have at the end of a year the even more imposing sum of $1.25 \cdot 1.25 \cdot 1.25 \cdot 1.25 = 2.44$ euros. She begins wondering how well she might do if she played this game of deposit and withdrawal daily, or hourly, or even every second.

Isabella was surprised to discover that a greater and greater frequency of transactions does not lead to arbitrarily large financial gain. There is a limit to how much interest can be earned. The limiting factor by which the initial deposit can be multiplied is the famous number $e = 2.7182\ldots$.

Just as for the layperson the numbers $1, 2, 3, \ldots, 9$ are the stuff of everyday encounters, for the mathematician, the number e pops up in

[1] We are going to assume that as a courtesy to customers who withdraw early, the bank gives the appropriate fraction of the interest earned.

the most amazing variety of places. Together with π it is one of the most important numbers in all of mathematics. You will always find it lurking when the subject is exponential growth (think bacteria) or exponential decay (uranium 235). However, it also appears frequently in probability theory, where it appears in the formula for the famous bell curve.

How Much Interest Is to Be Expected?

The following table demonstrates the surprising fact that while an increasing frequency of compounding of interest leads to a greater accumulation of interest, there is a limit to how much can be earned. The first row shows the number of interest periods in the year, that is, the number of times interest is compounded, while the second line shows the account balance at the end of the year, where we continue to assume an impressive interest rate of 100%.

Number n of Compoundings	Balance at Year's End
1	2.000
2	2.250
5	2.488
10	2.594
50	2.692
100	2.705

As n gets larger and larger, the balance at the end of the year approaches the number $e = 2.718281828459045\ldots$.

The Exponential Function

There is another way of getting at the number e. In formulating a simple model of population growth, one is led to the search for a function with the following properties:

- The function f should have a prescribed value at 0. With a suitable normalization, that value can be 1.
- The function should be differentiable. This means that one can speak unambiguously about the rate of growth of the function at each of its points. Its graph should have no "kinks."

- If the rate of growth at a point x is denoted by $f'(x)$, then it should always be true that

$$f'(x) = f(x).$$

This means in particular that the larger the value of the function, the faster it grows. (See Figure 1.)

The relation to models of growth should be clear, since as a population grows, the rate of growth can be expected to increase as well, since a larger population will have more reproducing pairs than a smaller one.

Remarkably, there is only one function that possesses all of these properties, namely the function that associates to a number x the value e^x. Therefore, one may introduce the number e as follows:

- In a first step, establish that the function f described above is in fact uniquely determined.
- Then define e as the value of this function at the point $x = 1$. Since $f(1) = e^1 = e$, we indeed get the desired number.

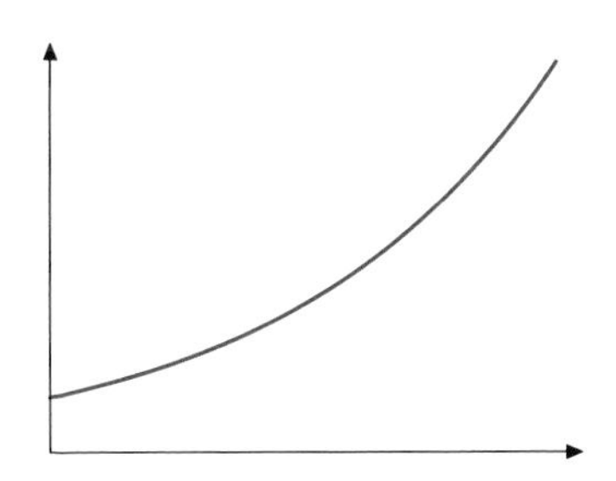

Figure 1. The exponential function.

The advantage of this method is that one immediately finds oneself in an important application area of the number e. Whenever one is modeling growth or decay processes (bacteria, nuclear fission, etc.), functions of the form e^{ax} appear. Here a is a positive number when the population is growing, and negative when it is declining.

Figures 2 and 3 give two typical examples.

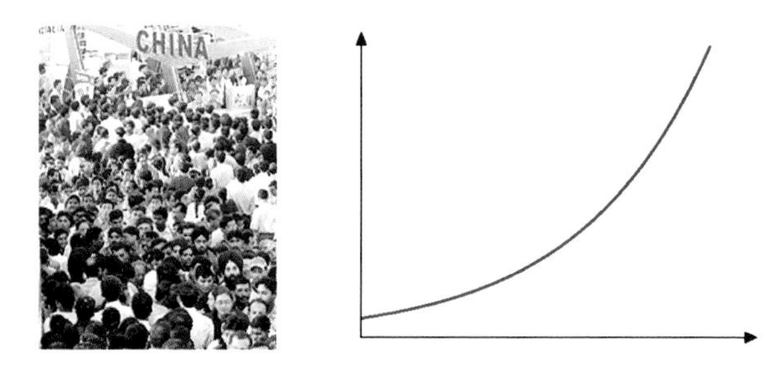

Figure 2. The function e^{ax} with population growth.

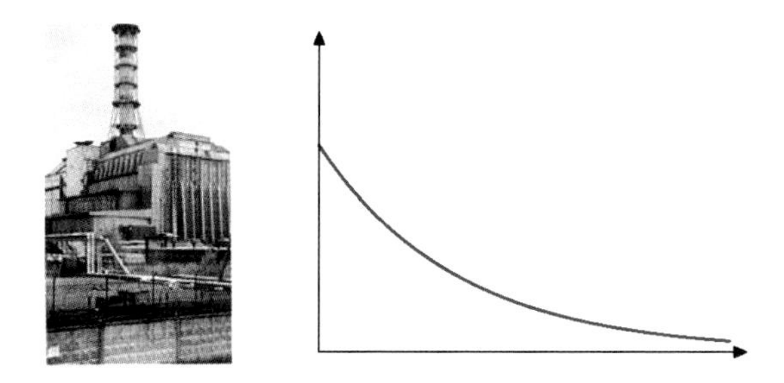

Figure 3. The function e^{ax} with radioactive decay.

In the first example, the size of the population (say the number of residents in a given country) is graphed as a function of time, while in the second, the amount of radioactive substance in a radioactively contaminated building is modeled.

Chapter 43

How Do Quanta Compute?

Several years ago there was a great deal of talk about quantum computers, though lately all has been rather quiet on that front. That is because while such computers could perform unbelievable feats of computation if the necessary complexity could be built into them, it seems at present that one should be rather pessimistic about their ever coming into existence.

Meanwhile, there has been intensive research into the capabilities of quantum computation. The situation is similar to that in the previous century, when before the launch of the first space rocket, much thought was given to what might be accomplished if space travel ever got off the ground.

The idea behind quantum computers is to exploit certain laws of the nanoworld that differ dramatically from those of our everyday experience. In particular, quantum mechanics teaches us that in the interaction of quantum systems, the probabilities of what is measured at the end superimpose in a controllable way. If one can transform a mathematical problem so that the solution could be represented on a quantum computer, then it would be possible to treat these superimpositions in parallel as a gigantic number of cases, in which the number of possibilities grows exponentially with the number of building blocks, the so-called quantum bits, or "qubits."

Unfortunately, there are many problems that make the possibility of constructing such a quantum computer doubtful. Some are of a physical nature. For example, the most notable properties of the quantum world could be exploited only if the system were extremely well shielded, since any stray particle, such as from cosmic radiation, could cause the system to crash. Furthermore, there are seemingly insurmountable problems with programming: if an intermediate calculation requires a certain result, that result first has to be codified. However, in the quantum world, every measurement alters the state of the system, and the initial state cannot be reconstituted. The result is that there are comparatively few interesting mathematical questions that can be treated by this method. Usually in mathematics one requires exact solutions, not those that are true only with a certain probability.

One example in which one can just keep trying for a solution is in the decryption of secret codes. And indeed, the interest in quantum computers arose when the American Peter Shor (shown in the photograph) devised a quantum algorithm for factorization of large integers, which would assist in cracking the RSA code.[1] For his work he was awarded the Nevanlinna Prize at the 1998 International Congress of Mathematicians in Berlin.

What Are Qubits?

The most important concept in connection with quantum computation is that of the quantum bit, or qubit. The term is meant to be reminiscent of the "bits" of "everyday" computers, where a bit is a unit of information storage that can assume one of two values, thought of as zero or one. Billions of bits are linked together for carrying out complex calculations.

[1]See Chapter 7 for more on public key cryptography, and Chapter 23 for details on the RSA algorithm.

A qubit is then the quantum-computation analogue of the bit. One can imagine a qubit as a black box that on receiving a query replies with a zero or a one. What is known are the probabilities with which the box will return each of the two values. In this sense, a "classical" bit is a specialized qubit, namely a qubit about which one is certain whether the output will be a zero or a one.

This probabilistic definition reflects the fact that the nanoworld is governed by probability, not certainty. It is only through measurement that it is decided which of the possible values will be concretely realized.

However, the image of the qubit as a black box is inadequate for describing the interaction of multiple qubits. For a more refined picture we must imagine that the probability for the output of a one or a zero is determined by a pair of arrows in the plane: the square of the length of the arrow labeled 1 gives the probability of a query being answered with a one. For example, if the length of the arrow labeled 1 is 0.8, then the probability of a one is $0.8 \cdot 0.8 = 0.64$. Clearly, then, the probability of a zero is $1 - 0.64 = 0.36$, and the arrow labeled 0 will have length 0.6 (since $0.6 \cdot 0.6 = 0.36$). In the picture, a qubit is shown whose probabilities of zero and one are about equal, and so they may be thought of as representing a coin that is tossed to determine whether 0 or 1 is to be output.

A qubit has direction as well as length, and what makes things interesting is that the directions of the "zero" and "one" probabilities are independent. The interaction of two qubits is represented by their arrows being placed together as in vector addition.[2] Thus two qubits

[2] If this vector addition results in a pair of vectors whose squares do not add up to 1, the vectors are "normalized," that is, multiplied by a factor that again makes possible the interpretation of the vector lengths as probabilities. For example, if a vector sum results in a pair of qubits in which the arrows for 0 and 1 have lengths 0.3 and 0.4, so that the square sum is $0.3 \cdot 0.3 + 0.4 \cdot 0.4 = 0.25$, one has to multiply each vector by $1/\sqrt{0.25} = 1/0.5 = 2$. This normalization results in vectors with lengths 0.6 and 0.8, and indeed, $0.6 \cdot 0.6 + 0.8 \cdot 0.8 = 0.36 + 0.64 = 1$.

pointing in opposite directions, each with a high probability of a 1, can have a resultant qubit with a very small probability of returning 1.

The next picture presents an example. On the left are depicted two qubits, where to avoid clutter we have placed the 0 arrow above the 1 arrow and suppressed the labels. Since the bottom (1) arrows are larger than the upper (0) arrows, each of these two qubits has higher probability of returning a 1 than a 0. However, when these qubits are added, the resulting qubit (shown to the right of the equal sign) will almost certainly return a zero.

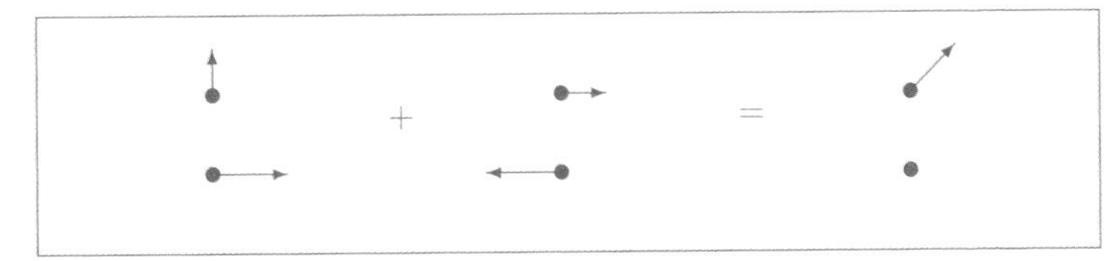

This principle is what underlies the (still hypothetical) functioning of a quantum computer. To answer a particular question, one queries a properly programmed quantum computer. In theory, a gigantic quantity of possible outputs will be produced, but the probabilities of the individual results are set in such a way that the desired result has a particularly high probability of being selected. This is all highly technical and complicated, and such problems are still a long way from being solved by quantum techniques.

The gigantic orders of magnitude involved come from the interactions of the qubits. Suppose we have two of them, Q_1 and Q_2. Each can take on the value 0 or 1, and so together the possible results are 00, 01, 10, and 11. If Q_1 and Q_2 are imagined as a quantum-mechanical system, there are *four* associated probability arrows. For example, if the 00 arrow is particularly short, this means that there is only a very small probability that the two qubits will be found together in state 0. In realistic cryptographic applications one would need several thousand qubits (resulting in a number of states on the order of a number with several *thousand* digits). This vastly exceeds what today is technically possible.

Chapter 44

Extremes!

What rotation speed leads to the best performance of a motor? How should a ski jump be constructed to ensure the longest jumps? Over the centuries, a host of methods have been developed to answer such questions of "biggest" or "smallest."

The simplest kind of problem involves a finite, relatively small, number of choices. Then one can simply try out all possibilities and choose the most favorable. Things get more complicated when one has to optimize a number of continuously varying parameters, for example in studying the distance a ball travels as a function of the angle at which it is thrown.

Readers who have had a course in calculus may recall having encountered such problems, which were solved by taking the derivative of the relevant function, setting it equal to zero, and solving the equation for the desired parameter. Observe that this method allows a problem that has an infinity of possible solutions (since the parameter varies continuously) to be solved by the solution of a single equation and is therefore treatable in a finite amount of time. This astounding fact was noted several centuries ago and was a primary motivation for the development of differential and integral calculus.

Things are not much different when several parameters are considered simultaneously. Again, everything is reduced to the solution of equations (though they become much more complicated). It should also be clear that with today's high-speed computers, problems of much greater complexity can be treated than was possible a few decades ago.

But sometimes an entirely new idea comes along. Several years ago, *simulated annealing* was all the rage.[1] Imagine a hiker overtaken by fog seeking the highest point around. He walks, heading upward whenever possible, though to avoid the problem of ending up at the top of a small hill instead of at the highest point, he sometimes goes downward as well.

Finally, it should be noted that one can delegate to mathematics the solution of problems, but not the setting of goals. The solution of motor optimization in our first example depends on whether one is seeking the most powerful, the most economical, or the most environmentally friendly motor.

A Typical Extreme Value Problem

Our friend Ferdinand is planning a bicycle tour through the Harz Mountains. He leaves his hotel in the morning and returns at night. It should be clear that at the highest point of his trip, his bicycle is exactly horizontal. For if the front wheel were higher than the rear wheel, there must be a higher point ahead, and if it were lower, he must have come from a higher point.

It is just such an idea that underlies extreme value problems. If one has reached a maximum, then the steepness (or *slope*) of the curve must be zero. Using the terminology of Chapter 13, one could say that a zero slope is a *necessary* condition for the existence of an extreme value.

To use this knowledge for the actual calculation of extreme values, one requires formulas to express the slope of a curve. This was one of the strongest impulses in the development of modern mathematics, namely the calculus, invented independently by Leibniz and Newton.

Here is an example: where does the function $-x^2+6x+10$ assume its greatest value? Figure 1 shows that the function's graph first rises and then falls.

---

[1] See also Chapter 60.

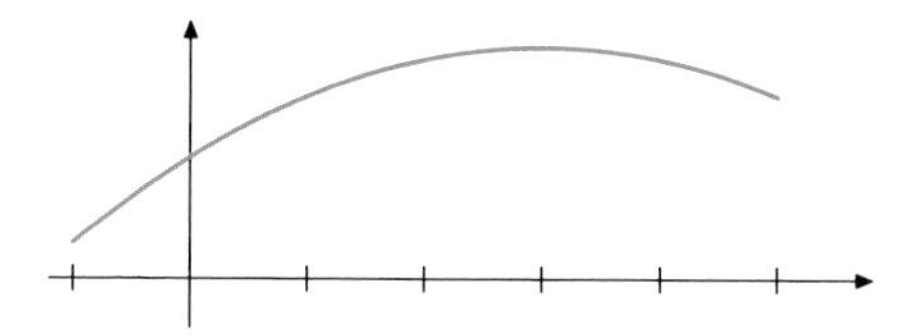

Figure 1. The function $-x^2 + 6x + 10$.

But where does it reach its maximum? Using the rules for the derivative, which we shall not derive here, one knows that the slope at a point x is given by $-2x + 6$. And this expression vanishes for $x = 3$. Therefore, the maximum value is achieved at $x = 3$. (One must be careful in solving such problems that one has not accidentally found a minimum of the function. After all, Ferdinand's bicycle was horizontal at the lowest point of his trip as well.)

Chapter 45

Infinitely Small?

For several centuries, infinitely small quantities drifted about the mathematical landscape. They terrorized all those who wished to put all of mathematics on as firm a foundation as those enjoyed by geometry and algebra.

These infinitesimal quantities first saw the light of day in the seventeenth century, when they were needed in the newly developed integral and differential calculus. There were two approaches, developed independently by Newton and Leibniz, and neither of them could get by without the notion of the "infinitely small."

But what was that supposed to mean? If a number x is positive, then there must be a smaller positive number (for example, $\frac{1}{2} \cdot x$). Therefore, there cannot be a smallest positive number. However, one can be misled in considering a quantity by passing to smaller and smaller scales of measurement.

For example, consider the arc of a circle. If we fix a point on this arc and view the segment of the arc surrounding the point under greater and greater magnification, the segment will look more and more like a straight line, and one is tempted to see the segment in the limiting case as an ordinary straight line. Then one could say that at the level of the infinitely small, a circle *is* a line.

That is how Leibniz argued when he passed from curves to their tangent lines; yet despite such dubious argumentation, he was able to draw many useful and interesting conclusions. Many of his mathematical contemporaries were critical of these developments, and it was not until the nineteenth century that the foundations of the calculus were developed to the point that one could finally jettison the idea of infinitely small (and infinitely large) quantities. In all this, an important role was played by the Berlin mathematician (please allow the author, a Berliner, to express pride of local patriotism) Karl Theodor Wilhelm Weierstrass.

No one mourned the disappearance of those infinitely small quantities. In particular, mathematical neophytes have an easier time acquiring a firm background in calculus without having to deal with such vague concepts. It is unlikely that there will be an infinitesimal renaissance. However, a few decades ago, an attempt at such a renaissance was made, under the rubric *nonstandard analysis*. But if one wishes to develop the concepts precisely, nonstandard analysis is much more difficult than any other approach to the secrets of differentiation and integration.

The World of Epsilon

How, then, does mathematics today deal with the infinitely small? As an example, let us consider the reciprocals of the natural numbers, that is, $1, \frac{1}{2}, \frac{1}{3}, \ldots$. Intuitively, it is clear that these reciprocals become "arbitrarily small" or "approach zero arbitrarily closely."

During Leibniz's time one would have said that reciprocals "eventually become zero," a turn of phrase that today would earn a rebuke if uttered even by a college freshman. The generally accepted way of making precise the idea of the infinitesimal goes like this (caution: things are about to get technical).

Meet ε

ε If one has a sequence of positive numbers $x_1, x_2, x_3, \ldots$, one says that they *converge to zero* if the numbers eventually become smaller than any positive number, no matter how small. More precisely, no matter what number ε (this is the Greek letter "epsilon") is given, no matter how small, there is some index n such that not only x_n, but also x_{n+1}, x_{n+2}, and all subsequent numbers in the sequence are less than ε. To establish this, one has only to find a way of producing an appropriate value of n for any given ε.

In our example, we could proceed as follows: Given a value of ε, we need a value of n such that n is larger than $1/\varepsilon$. If $\varepsilon = 1/1{,}000$, for example, one could choose $n = 1{,}001$. From the rules for evaluating inequalities, it follows that $1/n$ (and even more so $1/(n+1)$, $1/(n+2)$, etc.) is smaller than ε. We have thus established the correctness of the assertion "the reciprocals of the natural numbers converge to zero."

Admittedly, on first acquaintance this definition is somewhat difficult to digest. This is true for students of mathematics, no matter where they are studying, who have to internalize it in their first semester of calculus. What is important in this definition is that it makes precise an otherwise vague concept of fundamental importance so that one can formulate further mathematical statements with precision.

Nonstandard Analysis

In the area of mathematics called *nonstandard analysis*, which was developed in the 1960s, one imagines the numbers in such a way that every "classical" number is surrounded by a "cloud" of other numbers that are immeasurably close. The numbers that are close to the classical zero are called *infinitesimals*.

In this new extended domain of numbers, the usual rules hold: one can add and multiply, order of addition is irrelevant, and so on. The only thing that one must do is to accustom oneself to some strange properties of the notions of bigger and smaller: it is no longer true that every number is exceeded by one of the numbers $1, 2, 3, \ldots$.

Once one has become accustomed to this new world of numbers, many things that are usually difficult for beginners to grasp become very simple. For example, the slope of a function is no longer, as is usual today, a limit, but simply the relationship between the adjacent and opposite sides of an infinitesimal triangle, just as it was conceived by Leibniz.

However, despite these advantages, nonstandard analysis will remain a footnote in the history of mathematics. In order to understand how this approach is based on a firm axiomatic foundation, one needs several years' worth of study. But numbers and their properties can't wait that long: they are needed in the first weeks of the first semester.

Chapter 46

Mathematical Observations at the Fire Department

Today we shall once again consider how real-life experience can be modeled mathematically. This time we are going to look at the problem of avoiding incorrect decisions by establishing a procedure for choosing among possible responses to a given situation.

The typical textbook example of a decision procedure involves a fire department operator who receives a call that the local school is on fire. From the sound of his voice, the caller appears to be drunk. How should the fire department respond? Continue with their poker game at the risk that the school burns down? Or send out four fire engines, even though the call was likely a hoax?

The general abstract background to this problem is that in perceiving the world, one may make two types of error: (1) What one perceives is correct, but one rejects the conclusion. (2) One accepts a hypothesis as true, although in fact it is false. (Mathematicians call these type-1 and type-2 errors.)

This may sound rather abstract, but one encounters situations every day in the newspapers and in life in which such errors must be avoided. Should one ignore a red light late at night (assumption: no police around)? Would it be a good idea to approach the cute young woman at the disco who has arrived with a suspicious-looking character (assumption: it is only her brother)?

Evolution has taught us to evaluate such situations in split seconds. But these evaluations can vary greatly depending on one's character and life experience.

In the field of statistics, the proper evaluation of error forms the basis for decision procedures. Mathematics cannot prevent these two types of error from occurring, but one can attempt to quantify the consequences of a decision so as to minimize risk. And that is why the fire department responds to every alarm, even when the operator is "certain" that it is a false one.

Free-for-All at the Theater

The fire department example is a standard illustration of the two types of error. But since most of us have been involved neither in a fire nor in a false alarm, the example is perhaps too abstract and remote.

Therefore, we would like to emphasize that one meets with errors of types 1 and 2 every day in the newspaper. For example, a few weeks ago, a story appeared in which the Berlin police department moved in on the Deutsches Theater because they believed that a mass riot was about to begin. In fact, the only disturbance was that an inebriated theater patron had barged into another theatergoer. The press commentary was brutal: "Police state! Haven't they anything better to do?" Since the hypothesis of a riot was incorrect, the police were guilty of a type-2 error. What would the press have written, one wonders, if the error had been of type 1? "People were attacking each other, and again the police looked the other way!"

An even more dramatic example appeared in the Berlin *Daily Mirror* on 10 April 2006:

> *Emergency Responders Take Five-Year-Old's Call as a Joke*
> `Because a five-year-old's call to 911 was taken as a joke by the emergency operators, the boy's mother died. He had called the emergency number when his mother lost consciousness. The desperate lad was told to stop playing with the telephone. When help finally arrived, the boy's mother was dead.`

Even difficult personal decisions are subject to the two types of error. With respect to the hypothesis "I should have an annual checkup," an error of type 1 would be to skip the checkup with the idea that it might actually do more harm than good, when in fact, if you had gone to the doctor, an illness could have been detected at an early stage and cured. A type-2 error would have been made if one went for a checkup even though one was completely healthy and had nothing to show for it but a loss of time and money.

Chapter 47

The First Mathematical Proof Is 2,500 Years Old

When, actually, did mathematics begin? That is a tough question. It all depends on what one means by mathematics. If one means the ability to deal with simple problems related to numbers, then the beginnings lie in the dark recesses of prehistory. By the time of the Babylonians and Egyptians, calculation had reached a sophisticated level. How much wheat was harvested? How long must a ramp for pyramid construction be?

Those civilization had instructions for the calculations necessary to answer such questions, and they knew acceptable approximations to the number π and that right angles are related to what is today called the Pythagorean theorem.

Historians of mathematics generally set its beginnings at about the middle of the first millennium B.C.E. At that time, Greek mathematicians were no longer content with rules of thumb and example calculations. They wanted to get to the bottom of things, to establish a philosophical basis for truth. That is when the first mathematical proofs were developed, of which an early well-known example is Thales' theorem: if the vertex of a triangle is located on a semicircle such that the diameter of the circle is the longest side of the triangle, the triangle is a right triangle. (See Figure 1.) And this is always the case: one can give a rigorous proof based on simple assumptions.

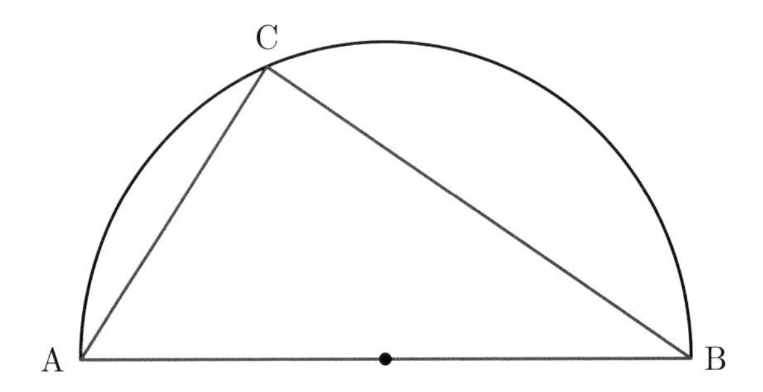

Figure 1. Thales' theorem: the angle at C measures 90°.

The kind of reasoning involved in the statement and proof of Thales' theorem reached a high point with Euclid's *Elements*, which collected all the knowledge of geometry known at the time and set forth a model for the development of a science that has since been frequently imitated. One begins with obviously true assertions (axioms) and then develops the rest with rigorous logic. It is thus, for example, that Newtonian physics is constructed, and even Kant considered the approach a model worthy of emulation: "In any study of nature, there is science present only to the extent to which mathematics can be applied to it" (I. Kant, *Critique of Pure Reason*).

This "provable search for truth" that was first realized by the Greek mathematicians led to remarkable successes. In fact, in recent years it has turned out that more and more phenomena in our environment can be described by facts that had earlier been discovered by mathematicians. In the case of Newton, the relationship was relatively simple: all that was needed were vectors and functions. Today, however, the experts can't get along without curved spaces, tensors, and probability distributions.

Why this should be the case is open to discussion. Was God Almighty a mathematician? Or are we capable of understanding only what is amenable to the methods that we have at our disposal? For mathematicians, such questions are of only secondary interest. They find it fascinating and reassuring to discover truths that will remain true forever.

On Semicircles and Right Angles

Thales' theorem is a good example of how a mathematical truth can be verified easily if one looks at the situation from the proper angle. Here again is the statement of the theorem (see Figure 1): Consider a semicircle lying above the circle's diameter. The endpoints of the diameter will be denoted by A and B. If C is then an arbitrary point on the semicircle, then the triangle ABC is a right triangle with its right angle at C.

One begins the proof by drawing an auxiliary line from the midpoint, M, of the circle to the point C, as shown in Figure 2.

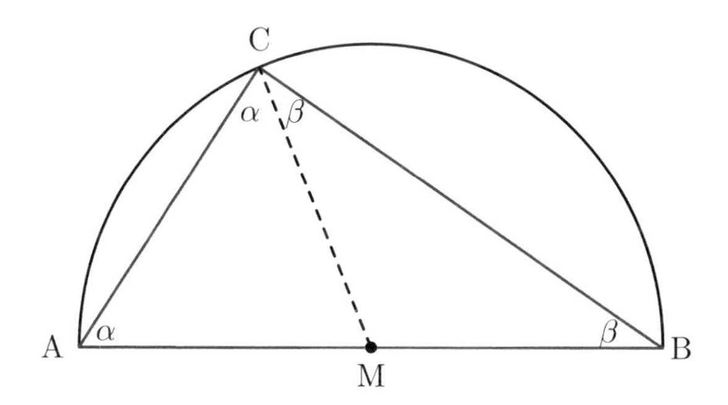

Figure 2. The proof of Thales' theorem.

Triangle AMC has two equal sides, since both AM and MC are radii of the circle. Therefore, in triangle AMC, the angle at A is equal to the angle at C. And likewise for triangle MBC: the angle at B is equal to the angle at C. Therefore, using the notation in the figure, the angle at C in the original triangle ABC is equal to $\alpha + \beta$. Now, it is known that in every triangle, the angle sum is equal to 180°. Therefore in our case, the angle at A plus the angle at B plus the angle at C must add up to 180°; that is,

$$\alpha + \beta + (\alpha + \beta) = 180,$$

which simplifies to $2 \cdot (\alpha + \beta) = 180$, or $\alpha + \beta = 90$. Therefore, the angle at C has measure 90°, as asserted.

Thales' theorem has wide application. For example, one can use it to prove the fact that straightedge and compass can be used to construct the square root of a number. This is explained in detail in Chapter 33.

Chapter 48

There Is Transcendence in Mathematics, Though It Has Nothing to Do with Mysticism

Mathematicians frequently borrow terms from other fields and apply them in ways that have little or nothing to do with their original meaning. This often leaves outsiders a bit puzzled.

Thus it is frequently supposed that transcendental numbers have something to do with mysticism and numerological secrets. Certainly many nonmathematicians owe their fascination with the number π to the fact that it is transcendental.

To understand what transcendental numbers are, one must understand some basic concepts about the hierarchy of numbers. For our purposes it suffices to begin with fractions, such as $\frac{3}{8}$ and $-\frac{7}{19}$. Such numbers are called *rational*, where the term derives from "ratio," and not from "reason."

Rational numbers suffice for solving most of the problems of daily life. But numbers such as π and square roots become necessary if one is interested in precise mathematical theory.

Numbers that are not rational are called, not surprisingly, *irrational*. Mathematics is full of such numbers. Among the irrational numbers are some that are particularly easy to describe. Such numbers are called *algebraic*. From a naive point of view, let us say provisionally that these numbers are somehow associated with algebraic operations such as addition, subtraction, multiplication, and division.

The numbers that are not algebraic are called *transcendental*. If one wishes to work with such numbers, one can no longer restrict oneself to algebraic methods. Such numbers often appear as limiting values in one or another mathematical construction.

And so what? The detailed study of the number hierarchy has led to spectacular results. The most famous of these is surely the proof of the impossibility of squaring the circle. The proof depends on the fact that only relatively simple (namely, certain algebraic) numbers can be constructed with straightedge and compass, but that squaring the circle requires the construction of a transcendental number. Since even the existence of transcendental numbers was not proved until the nineteenth century, it is no surprise that this problem remained open for two thousand years.

The Hierarchy of Numbers

Transcendental numbers are the most complicated members of the *hierarchy of numbers*, which plays a role in a number of the chapters in this book. Here we present a systematic overview.

Natural Numbers. These are the simplest numbers, $1, 2, 3, \ldots$. At some point in childhood, the abstraction "number" becomes internalized, and even preschoolers can master simple numerical calculations.

Worth knowing:

(1) For an axiomatic development of the natural numbers, one usually begins today with the *Peano axioms*, which establish that the set of natural numbers has an initial member and that one can "always continue to count." Of fundamental importance is the induction axiom: any statement that is true for the value 1 and for which one can prove that its

validity for n implies its validity for $n+1$ must hold for all natural numbers. (See Chapter 34.)

(2) The set of natural numbers is usually denoted by $\mathbb{N}$ (N is for "natural").

Integers. If one considers all possible differences of natural numbers, one obtains the *integers*. Thus the numbers 3, 0, and -12 are integers, since they can be obtained, for example, as the differences $5-2$, $4-4$, and $2-14$. The integers are useful for simple calculations in business, since, for example, they allow an account ledger to record both debits (negative numbers) and credits (positive numbers).

Worth knowing:

(1) The integers are usually denoted by $\mathbb{Z}$ (Z is for the German word for number, *Zahl*).

(2) Every natural number is also an integer, though the converse is false.

(3) Sums, products, and differences of integers are again integers. This is not true for quotients: while $44/11$ is an integer, $3/2$ is not.

Rational Numbers. A number is said to be *rational* if it can be written as a quotient m/n, where m in an integer and n a natural number. Examples: $33/12$ and $-1111/44$.

Worth knowing:

(1) The universal symbol for the rational numbers is $\mathbb{Q}$ (Q is for "quotient").

(2) If m is an integer, then it can be (somewhat artificially) written as $m/1$. Therefore integers are rational.

Irrational Numbers. Numbers that are not rational are called *irrational*. It came as a shock to the mathematicians of ancient Greece when they discovered that such numbers exist. The best-known example of an irrational number is the square root of two, about which we shall have more to say in Chapter 56. There is no generally used symbol for the irrational numbers.

Algebraic Numbers. Let us begin by imagining a game. The first player, Ferdinand, chooses a number x, and the second player, Isabella, must attempt to produce the number zero from x using the natural numbers and the symbols $+, -, \cdot, \div$. The number x can appear any number of times. If she can give a suitable formula that gives zero, she wins. Otherwise, she loses.

Here are a few examples:

- Ferdinand chooses $x = 17$. Isabella can easily win by proposing the formula $x - 17 = 0$. Isabella can clearly always win if Ferdinand chooses a natural number.
- This time Ferdinand tries to stump Isabella with $x = 21/5$. But Isabella is no slouch. She simply produces the formula $5 \cdot x - 21 = 0$, which shows that x can be transformed to zero under the allowed rules. More generally, Isabella can win whenever x is a rational number.
- Now Ferdinand plays his trump card and offers $x = \sqrt{2}$. Isabella has to think for a moment, but then she realizes that $x \cdot x - 2 = 0$ does the trick. So $x = \sqrt{2}$ can also be transformed into zero.

Numbers x that can be so transformed are called *algebraic*. We have seen in the game between Ferdinand and Isabella that integers, fractions, and the square root of two are algebraic.

Transcendental Numbers. Finally, if one knows what an algebraic number is, it is easy to understand what a transcendental number is. Namely, a number is *transcendental* if it is not algebraic. That is, a number x is transcendental if Isabella (or anyone else) is unable to transform x into zero using the rules described above, no matter how complicated the formula.

Worth knowing: It is important to observe the difference between how one must prove that a number is algebraic and how one must prove that a number is transcendental. To prove a number algebraic, one must come up with a formula and show that it transforms the number to zero. To prove a number transcendental, on the other hand, one must show that zero never results from any formula at all, no matter how complicated, no matter how long, even if it stretches from here to the Sun. Clearly, nonexistence is much more difficult to prove then existence, and indeed, it took to the middle of the nineteenth century before a particular number was rigorously proved to be transcendental.

Some of the most important numbers in mathematics are transcendental. The most famous of these are the base of the natural logarithms e and the number π (compare Chapters 16 and 42).

Chapter 49

Is Every Even Number the Sum of Two Primes?

We have often written in this book about prime numbers, which are the numbers $2, 3, 5, 7, 11, \ldots$ that are divisible by only themselves and 1. Although these numbers are simple to describe, they are at the center of a number of difficult problems. One of these has remained unsolved for centuries: the Goldbach conjecture.

Christian Goldbach (1690–1764) was a diplomat with an interest in mathematics. In 1742, he communicated his problem to the great mathematician Euler. Goldbach's conjecture is easily stated. It involves an additive property of the prime numbers. *Is it true or false that every even number greater then 3 can be written as the sum of two prime numbers?* To see what this means, consider the even number 30. Indeed, it can be written as $23 + 7$, and both 23 and 7 are prime. There are, in fact, other ways of writing 30 as the sum of two primes, such as $11 + 19$. And so it is for every even number that has ever been checked: it can be written as the sum of two primes, and for large numbers there are many, many possibilities.

Because of this overwhelming experimental evidence it is considered a scandal in the world of mathematics that up to now, no proof of this conjecture has been found. To be sure, there is no immediate use for such a theorem in any area of applied mathematics. However, in this book it should be made clear that mathematicians do not invest their energies only in the development of methods that are important for applications, but also in the discovery of general laws in the world of numbers, geometry, and probabilities.

Mathematicians are also seduced by problems that have remained unsolved for a long period of time and that have stumped the cleverest minds in the past. Some motivation might also be found in the prospect of coming into some money, since a prize has been offered for the solution to this problem.

Is the Goldbach Conjecture Important?

There is disagreement among mathematicians about the importance of the Goldbach conjecture. It is of course of interest, since it has managed to withhold its secrets for several centuries. The exhilaration of solving it would be like that of the first climbers to conquer Mount Everest or the first runner to run one hundred meters in under ten seconds.

To understand the skepticism regarding the significance of the problem, it is important to keep in mind that prime numbers are defined by a *multiplicative* property: a prime number cannot be written as a product of smaller numbers. Moreover, the most important result about prime numbers is related to multiplication: every natural number[1] greater than 1 can be written as a product of prime numbers, and the prime numbers used in the product representation are uniquely determined. However, the Goldbach conjecture involves sums of prime numbers. Why, ask the critics, should that be of interest?

The Experimental Evidence

Figure 1 depicts a graph whose horizontal axis represents the even numbers $z = 2, 4, 6, \ldots$, and over every such number z a point appears that represents the number of ways that z can be represented as the sum of two primes. For example, the green point near the left-hand

[1] The numbers $1, 2, 3, \ldots$.

edge of the graph lies above the number 14. Its height above the horizontal axis is 2, since 14 can be written in two ways as the sum of two prime numbers, namely as $3 + 11$ and $7 + 7$.

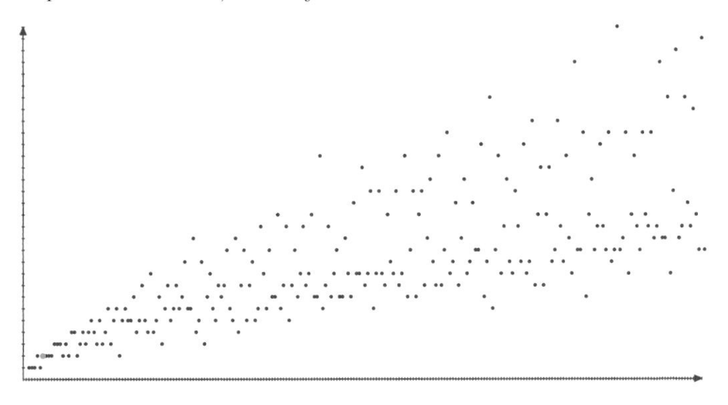

Figure 1. The Goldbach conjecture: the first 240 values.

The Goldbach conjecture is then equivalent to the assertion that there is no number z on the horizontal axis whose associated dot is at height zero. Based on the graph, one could in fact conjecture a great deal more. Even though the pattern seems to be a bit chaotic, it appears that even for the numbers z that have few representations as the sum of two primes—that is, the numbers whose associated dots lie along the bottom edge of the cloud of points—the graph moves steadily upward. In other words, not only should every even number be representable as the sum of two primes, but the number of different representations should become arbitrarily large for large z.

A "Proof" of the Goldbach Conjecture

The Goldbach conjecture is one of those famous problems that attract amateur mathematicians. A few weeks ago, a letter arrived at a university mathematics department containing the following "proof":

> First: There are infinitely many prime numbers.[2]
> Second: Therefore, infinitely many numbers result from adding pairs of prime numbers. This proves the Goldbach conjecture.

[2] See Chapter 4.

Unfortunately, that is scarcely a complete proof. The observation is certainly correct that it occurs infinitely often that a number is the sum of two primes, and of that fact there is no better proof. But that is nowhere near proving that *every* number can be so represented. Perhaps the correspondent had the following in mind: If I mark infinitely many elements of an infinite set, then that must be all of them. For finite sets the analogous assertion is correct: anyone who has five envelopes and sticks five stamps onto five different envelopes has certainly applied postage to all of the envelopes. However, the laws of infinite sets are different, and therefore, the Goldbach conjecture is still waiting for a correct proof (and the prize money that comes with it).

Chapter 50

Why We Invert Conditional Probabilities Incorrectly

Evolution has prepared us well to estimate probabilities. In split seconds we evaluate a situation and decide: fight or flight? try to extinguish the flames or run to safety? We are also adept at understanding the effect of new information on the probability of some occurrence. If, for example, you are wondering whether your new acquaintance is interested in classical music, you will likely decide that the odds are diminished if you discover that he confuses Schumann with Schubert.

These rather vague ideas can be made mathematically precise under the rubric of *conditional probability*. As a mathematical example let us consider the probability of rolling an even number on the throw of a single fair die. It is certainly $\frac{1}{2}$. However, if one has the information that the number rolled is a prime number, this probability sinks to $\frac{1}{3}$, since of the three prime numbers between 1 and 6, namely $2, 3, 5$, only one of them, 2, is even.

There is a mathematical formula, the renowned *Bayesian formula*, which allows conditional probabilities to be inverted. Imagine a bartender who knows from experience the percentage of customers who leave a tip. Say the average is 40%, and that among tourists, the average rises to 80%. Therefore, the information that a particular customer is a tourist increases the probability that he will leave a tip. The Bayesian formula allows a converse inference to be made: from the fact that a tip was left, one can compute the probability that the customer was a tourist.

Admittedly, customer tipping probabilities do not constitute a problem of fundamental importance. However, the same techniques are applicable to far more significant questions. A famous example is the efficacy of medical tests. What is the probability that I have a certain disease if the test for that disease comes up positive? To all those who may at some time experience such a positive result, mathematics can reassure them that the probability is much less than they might naively suppose. In this case, evolution has programmed us to be much too pessimistic.

The Measles Test

We had quite a bit to say on the topic of conditional probabilities and the Bayesian formula in Chapter 14 (the Monty Hall problem). Here we summarize the most important points:

- If A and B are two possible outcomes of a random experiment, then $P(A \mid B)$ denotes the probability that A occurs given that one knows that B has occurred. Here is an example: A card is drawn from a standard 52-card deck. Let A be the outcome "the jack of spades was drawn," and B the outcome "a spade was drawn." Therefore, the probability of A occurring is $1/52$, since there are 52 different cards, each with the same probability of being drawn. However, if one knows that a spade was drawn (and hence that B has occurred), then the probability of the jack of spades increases to $1/13$, since there are 13 different spade cards in the deck.
- The simplest form of the Bayesian formula involves two events, A and B. It is assumed that one knows $P(B)$, the probability of B occurring, and the numbers $P(A \mid B)$ and $P(A \mid \neg B)$.[1] The Bayesian formula then allows $P(B \mid A)$ to

[1] Here $\neg B$ stands for the complementary event to B (B does not occur). Thus if B is the event "spade," then $\neg B$ is the event "heart, diamond, or club."

be determined:

$$P(B \mid A) = \frac{P(A \mid B)P(B)}{P(A \mid B)P(B) + P(A \mid \neg B)(1 - P(B))}.$$

Now we can make our medical example more precise. Suppose we are concerned with the diagnosis of a rare disease. Let it not be cancer or AIDS. Suppose it is measles. One morning you observe a red pustule on your face and would like to know whether you have contracted measles. The doctor performs a measles test, and the result is positive. Do you have the disease or not?

For our analysis, let us suppose that A denotes the outcome "the measles test comes back positive," and B is the event "I have measles." To use the Bayesian formula, we need the numbers $P(B)$, $P(A \mid B)$, and $P(A \mid \neg B)$. The first of these is the probability of someone having measles. The disease is rare among adults, and we may set the probability at $P(B) = 0.05$, or 5%.

The probability $P(A \mid B)$ describes the reliability of the test: what is the probability that those who are sick with measles have a positive test? If the test were perfect, this probability would be 1.0, or 100%. However, there are no such tests, and one can only hope to approach the ideal. Let us optimistically set this probability at 0.98.

Finally, we need $P(A \mid \neg B)$: what is the probability that I have a positive measles test even though I don't have measles. Here one would hope for the answer to be zero, but such a goal is unachievable. A realistic probability of a "false positive" result is $P(A \mid \neg B) = 0.20$.

Now we can do the calculations. We would like to know $P(B \mid A)$, the probability that a positive test result implies that one has measles. Using the Bayesian formula, we obtain

$$\begin{aligned} P(B \mid A) &= \frac{P(A \mid B)P(B)}{P(A \mid B)P(B) + P(A \mid \neg B)(1 - P(B))} \\ &= \frac{0.98 \cdot 0.05}{0.98 \cdot 0.05 + 0.20 \cdot (1 - 0.05)} = 0.205\ldots. \end{aligned}$$

The probability of really being sick is a comforting twenty percent. This result is surprising. Most people would expect a higher number. The reason for this is that in estimating the probability, one tends to neglect the fact that the disease itself occurs only rarely.

A Geometric Visualization

To clarify why we fail at estimating these probabilities, consider the rectangle in the figure:

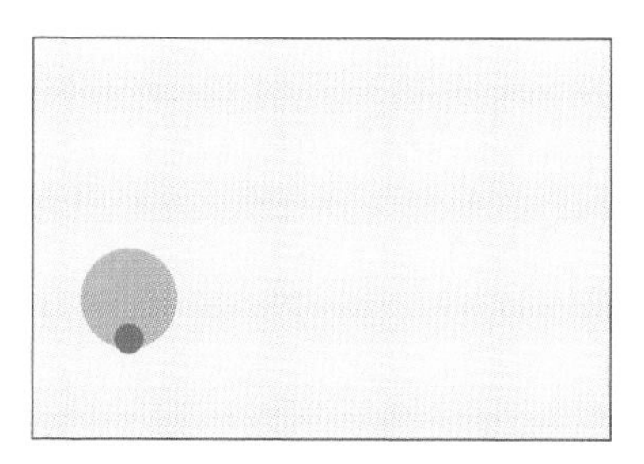

The rectangle symbolizes all the possible outcomes of interest. The small dark circle stands for outcome B: "measles." The circle is very small because measles is rare. The other circle represents outcome A: "test result positive." It cuts deeply into the small circle, since in the case of actual disease, the test almost always returns a positive result. The portion of the small circle outside the larger circle is small, since we are assuming a negligible number of "false negatives."

However, despite these conditions, the portion of the A circle taken by the B circle is not large: a positive test result does not mean that one almost certainly has measles.

Chapter 51

Millionaire or Billionaire?

Very large numbers are often to be found in the newspapers and other media: gross domestic product, national debt, and so on. We may assume that almost everyone knows that one million is represented by a one followed by six zeros. Without such knowledge, how could one realistically plan how to spend vast sums that one hopes to win in next week's lottery?

It is more difficult to comprehend numbers in the billions, perhaps because politicians and business executives don't really explain the significance of the large numbers that they throw around. For those who have forgotten, a billion is a thousand million. Thus a billionaire could give one-thousandth of her fortune to a pauper and turn him into a millionaire.

Just to make things confusing, there are different words in different languages for large numbers, and even worse, the same word can have different meanings even in the same language in different countries. In American English, and increasingly in England as well, the word "billion" represents what the English and the Germans call a "milliard." Thus in America, after million comes billion, then trillion, quadrillion, and so on. In German, milliards and billiards are squeezed in to muddy the picture, and one can easily become confused.[1]

Fortunately, here in Germany we at least have the good fortune that the terminology is stable. When one reads large numbers in British English or in French, for example, one had better see the data in addition to reading the words when one is reading older literature. Indeed, the two systems coexisted for a time. Today, the French have their *milliard*, like the Germans, and most Britishers now say "billion" along with their American counterparts. Therefore, be somewhat skeptical if you read in the German press that some American pop idol is a billionaire. In German reckoning that is "merely" a milliardaire. In fact, genuine quadrillions hardly ever appear, since even the gross domestic product can be measured in trillions. However, if one wished to write about the total amount of money in all the savings accounts in Germany, one would certainly summon the quadrillions.

Sometimes, mathematicians are asked the names for even bigger numbers. The answer is not very evocative, since larger numbers are generally expressed simply as powers of ten. Few mathematicians would speak of a billion to indicate the number represented by a one followed by nine zeros. They would simply refer to ten to the ninth power (10^9). And if ten to the thousandth power were to arise in conversation, one would not need to search in the Latin dictionary for the appropriate prefix.

What Difference Do Two Zeros Make?

It is a pity that nature failed to endow us with the capacity to grasp large quantities. One certainly has no trouble telling the difference between a gift of 10 euros and one of 1,000. But when one reads that light travels 5,870,000,000,000 miles in a year (almost six trillion), our capacity to grasp such a number fails. Two zeros more or less would make no impression on us one way or the other at this size. Such large numbers are simply "immeasurably large."

[1] Translator's note: A German *Milliarde* is an American billion, and a *Billiarde* is quadrillion, that is, a thousand million million: 1,000,000,000,000,000.

Unfortunately, one result of this incapacity is that some of the realities of everyday politics are difficult to understand. The announcement, "Berlin's deficit is 59,253,104,304 euros"[2] is certainly easier to make sense of if one thinks of it as "Berlin has a large deficit" than if one instead tries to digest the significance of about sixty billion. Nonetheless, it is sixty billion euros! If one had that kind of money, one could make millionaires out of everyone in a midsize city such as Herford. Or you could fill each of about six hundred boxes having a capacity of a cubic yard with hundred-euro notes.

We will certainly never live to see the day when everyone the world over will have the same notion of "billion." It's the same with driving on the left or railroad-track gauges. Once a society has become accustomed to a particular system over generations, there are thousands of good reasons not to change anything. Anyway, it causes problems mostly for journalists. Whenever the word "billion" arises, they have to figure out from what time and place the report was issued.

[2]From the Berlin newspaper *Tagesspiegel* of 22 March 2006.

Chapter 52

Mathematics and Chess

Can you play chess a bit? Do you at least know the rules? There are some aspects of mathematics that can be more easily explained by transferring them to some other realm. Today we are going to use the world of chess.

First let us consider the rules of the game. These correspond to the axioms of mathematics. There are no serious attempts being made to invent new rules for chess. Much more effort is devoted to figuring out how to use the existing rules to win games. Analogously, a mathematician labors for months, even years, to determine whether a particular result in some theory is provable.

Also, as everyone knows, the truths about chess are independent of particular chessboards and players. If there is a path to checkmate from a certain position, then it can be written down on a piece of paper (or even reported orally if necessary), and that is that. Likewise, mathematical results are not bound to individuals, books, or languages. It is indeed difficult to localize mathematics. Plato considered it eternal, located in the realm of ideas. Other philosophers saw it as simply the set of consequences to be drawn from conditions created by convention, which sometimes, incidentally, have useful applications.

Now we come to solved and unsolved problems. Even beginners quickly learn that the endgame can be won by a king and rook against an opponent who has only the king remaining. However, we will likely never learn whether at the beginning of the game white has a winning strategy. The game is just too complex. Similarly, there are many problems in mathematics that remain open, and no one knows whether and when one of them might be solved. (Some of these, such as the Goldbach conjecture, have been mentioned in this book.)

However, there is a fundamental difference between mathematics and chess, which explains why none of the major universities has a department of chess. With chess one cannot determine the stability of a bridge or calculate the odds of winning the lottery. In contrast to mathematics, chess is not directly applicable to the problems of the world. Why it is that "the book of nature is written in the language of mathematics" (Galileo), no one really knows.

How Should One Study Mathematics?

There are further parallels between mathematics and chess. Consider the university education of a mathematician. As a rule, one learns mathematics by solving concrete problems: "Prove that the number x is irrational." "Show that the given differential equation has an infinite-dimensional solution space."

That is similar to the solution of chess problems: "Black to move and win" (see Figure 1). "How can white gain an advantage by a bishop sacrifice?"

However, every chess player knows that one does not become a master through solving chess problems alone. When one is confronted by a chess position and is seeking a good move, one doesn't know whether there is a mate in three moves or whether a spectacular sacrifice might turn the tables.

The analogy in mathematics is that in one's education one must become involved in situations in which it is unclear at the outset what might turn out to be provable. Here creativity is required to judge what mathematical methods might be brought to bear on the question at hand. Such situations correspond much more closely to what a professional mathematician does at the office than conditioned responses to "show that...."

Figure 1. Black to move and win.

Chapter 53

"The Book of Nature Is Written in the Language of Mathematics"

"The book of nature is written in the language of mathematics," wrote Galileo almost four hundred years ago. By this he meant that many aspects of reality can be profitably translated into the language of mathematics. Imagine, for example, that you have decided to carpet the living room of your new home. Then you can estimate how much the carpeting that you have selected will cost with the help of some elementary geometry: simply calculate the area of a rectangle.

In figuring your costs, certain aspects of your real living room were translated into mathematical language. In applications of this notion to engineering and the natural sciences, the idea is the same: the aspects of interest in the problem are translated into the language of mathematics and then solved using mathematical techniques. The subsequent translation back to the real world should then—one hopes—solve the real-world problem. Almost every branch of mathematics is called on in this process: geometry and algebra, numerical methods and probability. The problems to be solved can be as complicated as you like.

In the end, this is all not much different from the situation in which a Berliner vacationing in the USA translates the question, "Wo ist die nächste Tankstelle?" into its English equivalent, "Where can I find a gas station around here?" and then hopes that the natives will be helpful. The solution will likely be rendered in English, which one can then translate back into the original language.

No one today seriously doubts that Galileo had it right. What is debated, though, is why this should be so. Is it a mystery whose secrets we are incapable of penetrating? Is God a mathematician? That is, is the world constructed according to mathematical principles that we are progressively able to decipher? Or is it all simply a matter of convention, and the applicability of mathematics is nothing but an illusion?

For many centuries, mathematicians and philosophers have been debating this question without a satisfactory resolution. There is little hope that they will ever find the answer.

The Mathematician as Translator

The application of mathematics to real-world problems is often described as translation: translate the fundamentals of problem P into problem P', find a solution S', and then formulate the reverse translation S of the solution as a (possible) solution of the original problem. This process is illustrated in Figure 1.

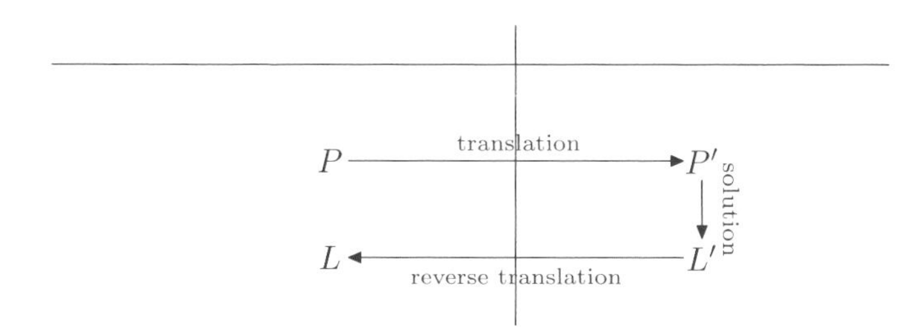

Figure 1. Mathematics as translation.

This all has great similarity to other techniques of translation within mathematics and from the real world. Thus in Chapter 36 we indicated that the most important advantage of calculating with logarithms was that multiplicative problems could be translated into additive problems. And a Belgian arriving at JFK airport in New York and looking for a taxi would do well to translate the problem into English so that an American can solve it.

The Danish mathematician Vagn Lundsgaard Hansen made an effective depiction of the role of mathematics as "bridge to the world" in a poster designed for the "World Mathematical Year 2000."

The poster shows the mile-long Storebælt Bridge, the longest suspension bridge in Europe—and in the world second only to the Akashi-Kaikyo Bridge, in Japan. It links the Danish islands Fyn and Sjælland.

You Don't Need Spherical Trigonometry to Cultivate Your Garden

We should note that in creating a mathematical model, one is forced to make certain simplifying assumptions. But if the model is too simple, then the results will be poorly applicable to the real-world situation being modeled, while if it too complicated, the computations will become too involved to be practicable or even impossible. No one employs spherical trigonometry in laying out a backyard garden. Mathematical theory is useful only if a suitable model is constructed.

It is also important to observe that translation into a mathematical model is only the beginning. When, for example, the braking distance of an automobile is to be determined, the laws of mechanics are also needed, since they give a relationship among masses, forces, and motion.

In complex situations it can happen that a large number of theories about the nature of the world are needed to arrive at a concrete mathematical problem. And if the solution fails to correspond to observation, it may not be clear which of the theories has to be modified.

Chapter 54

The Search for Mersenne Primes

Does your computer ever get bored? Would you like it to help get your name inscribed in the annals of mathematical history? Then surf your way over to the web site www.mersenne.org. There you will find a computer network devoted to the goal of finding very large prime numbers.

Recall that prime numbers are those that can be divided only by themselves and 1, numbers such as 3, 11, and 31. It is known that there are infinitely many primes, and therefore there are arbitrarily large ones. But that does not mean that one can necessarily produce examples of such large prime numbers. The problem of finding large primes is tackled from a variety of approaches, with a mixture of theoretical considerations and massive computer calculations having proved the most effective.

Naively, one could suppose that it would be easy and fast to prove whether a given number is prime. One must simply test each smaller number to see whether it is a factor. Unfortunately, such an approach is feasible only for relatively small numbers. As numbers get larger, this form of testing can quickly demand computation times of order the age of the universe.

Therefore, in searching for the world's largest known primes, only special candidates are considered. These numbers arise by multiplying the number 2 by itself a large number of times and then subtracting 1. For example, the numbers 31 and 63 result from $2 \cdot 2 \cdot 2 \cdot 2 \cdot 2 - 1$ and $2 \cdot 2 \cdot 2 \cdot 2 \cdot 2 \cdot 2 - 1$. Such numbers are called *Mersenne numbers*, and

if, as in the case of 31, the result is a prime, the number is called a *Mersenne prime*. They are named for Father Marin Mersenne (1588–1648), shown in the figure, a man who served science as well as the Church.

There is a test for whether a Mersenne number is a Mersenne prime that can be carried out in a reasonable amount of time, even for very large numbers. One has to perform a single divisibility test on a very large number. This is best performed on a large network of computers, and it is the details of coordinating such an arrangement that are organized by the Mersenne Network.

Every now and then, a new record-breaking Mersenne prime is discovered. The prime number champion for 2004 was discovered in November. It has over six million digits. A certain Michael Shafer had the good fortune that it was on his computer that the test returned a positive result. As a codiscover of the fortieth Mersenne prime he became as well known as many a professional mathematician.

Prime Number Records

Prime number records age fast. With ever-increasing computer speed, expanding computer networks, and more-refined algorithms, larger and larger primes are being discovered. It is therefore not surprising that the record announced when this newspaper column appeared in 2004 has been superseded. While these lines are being written, the number

$$2^{25,964,951} - 1$$

is the record holder. This could quickly change.[1] If you would like to know where things stand today, surf to www.mersenne.org, where the record books are kept.

To get an idea of the enormous size of these numbers, recall that the number 2^{10} equals 1,024. Put another way, 2^{10} is equal to about 10^3. Analogously, we may conclude that $2^{20} \approx 10^6$, $2^{30} \approx 10^9$, and so on. In general, 2^n corresponds approximately to a number that begins with 1 and is followed by $3 \cdot (n/10)$ zeros (at least when this fraction is a whole number). For our example $2^{25{,}964{,}951} - 1$, we should expect a number with $3 \cdot (25{,}964{,}951)/10$, that is, about eight million, digits. If we wanted to print such a number, then if every page could accommodate fifty lines of one hundred characters each—and therefore 5,000 digits per page—we would need $8{,}000{,}000/5{,}000 = 8{,}000/5 = 1{,}600$ pages. That would be quite a hefty tome.

Primality Tests

How can one quickly determine whether a number n is prime? For example, is 2,403,200,604,587 prime?

The most naive course is to test all smaller numbers m to determine whether m is a factor of n. This requires roughly n calculations, which for large numbers would take much too long.

We could save some time with a little thought. Namely, if n is not a prime number and thus can be written $n = k \cdot \ell$, then k and ℓ cannot both be larger than the square root of n. (From $k > \sqrt{n}$ and $\ell > \sqrt{N}$ we would conclude by multiplication that $k \cdot \ell > \sqrt{n} \cdot \sqrt{n} = n$.) Therefore, if there is no divisor in the range 2 to $\sqrt{n}$, then n must be prime.

The savings in effort are dramatic. For a number of order of magnitude one million, we would need only about one thousand tests instead of one million. However, if n is a number with several hundred digits, the savings are still insufficient to make the method usable: $\sqrt{n}$ would be so large that it would take several hundred years to finish the checking.

[1] Translator's note. And indeed it has. Since then, two new Mersenne primes have been discovered, the most recent in September 2006. The current record holder, $2^{32{,}582{,}657} - 1$, has close to ten million digits.

Therefore, another way must be found. A procedure that works for discovering new record holders is available only for numbers of the type $2^k - 1$. It is called the *Lucas–Lehmer test*:

> Let us introduce the abbreviation $M_k = 2^k - 1$. When is M_k a prime number? It can be proved that this can be the case only if k is itself a prime, although that does not mean that M_k is necessarily prime. (For example, $M_{11} = 2^{11} - 1 = 2{,}047 = 23 \cdot 89$.)
> One therefore chooses a prime number k and defines numbers $L_1, L_2, \ldots, L_k$, called *Lucas–Lehmer numbers*, as follows: $L_1 := 4$, $L_2 := L_1^2 - 2 = 14$, $L_3 := L_2^2 - 2 = 194$, and so on. We always have $L_{\ell+1} = L_\ell^2 - 2$. Then M_k is a prime number precisely when M_k is a divisor of L_{k-1}.

To see how this works, let us compute the first few Lucas–Lehmer numbers:

$$4, \quad 14, \quad 194, \quad 37{,}634, \quad 1{,}416{,}317{,}954, \quad \ldots .$$

It is clear that the numbers grow very rapidly. However, we are concerned only about whether these numbers are divisible by M_k, and therefore it suffices to consider L_ℓ modulo M_k.[2]

Example 1. We set $k = 5$. Then $M_k = 2^5 - 1 = 31$. Since 31 is prime, the test should be positive. We must determine the numbers L_1, L_2, L_3, L_4, and check whether the last of these is divisible by 31. Calculating these numbers modulo 31, we obtain $4, 14, 8, 0$, and since the last result was zero, the test correctly informs us that 31 is prime.

Example 2. Now let us investigate $k = 11$. We need to test $M_{11} = 2{,}047$. Here are $L_1, L_2, \ldots, L_{10}$ modulo 2,047:

$$4, \ 14, \ 194, \ 788, \ 701, \ 119, \ 1{,}877, \ 240, \ 282, \ 1{,}736.$$

Since the last of these is not zero, M_{11} cannot be prime. (By the way, note that this method proves that M_{11} is not prime, but it does not produce a divisor. That is a feature that this procedure does not offer.)

[2]See Chapter 22 for a discussion of modular arithmetic.

Chapter 55

Berlin, Eighteenth Century: A Beautiful Formula Is Discovered

A few years ago mathematicians were polled to determine an answer to the question, "what is the most beautiful formula?" Formulas were proposed from many different areas of mathematics. The formula that won was discovered by the Swiss mathematician Leonhard Euler in the eighteenth century. At the time, Euler was court mathematician to Frederick the Great in Berlin.

To understand the formula, you will need to recall a few of the most important numbers in mathematics. These include, of course, zero and one, since with these one can construct all the remaining numbers. Furthermore, their properties are essential in operating with numbers, because zero is the additive identity—that is, adding zero to a number leaves the number unchanged—and analogously, 1 is the multiplicative identity: for any number x, we have $1 \cdot x = x$.

Then one needs the number π. Even schoolchildren know this number from calculating the area and circumference of a circle from its radius. And for describing certain kinds of growth phenomena, the number $e = 2.71828\ldots$ is essential. Exponential growth (bacteria) and exponential decay (radioactivity) belong to the foundations of mathematical modeling, and in both cases, the number e appears. Finally, it has been clear for several centuries that the set of numbers necessary for solving algebraic equations has to be enlarged to include the *complex numbers*, which is accomplished by defining the number i as an "imaginary" square root of -1. These numbers are useful not only in theoretical investigations. Indeed, complex numbers belong in many toolboxes, for example that of an electrical engineer.

$$0=1+e^{i\pi}$$

Remarkably, there is an intimate relationship among 0, 1, π, e, and i. Namely, if to 1 you add e raised to the power i times π, that is, the quantity $1+e^{i\pi}$, you end up with zero. That is Euler's formula.

For mathematicians this formula has special significance, since it highlights the unity of all mathematics. After all, it is somewhat mysterious that a bunch of numbers that were custom built for a variety of purposes should have such a simple relationship.

The Most Beautiful Formula: The Proof

The chapters in this book give prominent play to almost all of the numbers that appear in Euler's formula: Chapter 16 for π, Chapter 28 for zero, Chapter 42 for e, and Chapter 94 for i. How did Euler discover this formula?

To understand the formula, one needs to know a few mathematical functions. An important role is played by the fact that sometimes, complicated expressions can be well approximated by simple sums. For example, if a number x is "small enough," the square root of $1+x$ can be approximated by $1+x/2$. Let us check this for the small value $x=0.02$: we have $\sqrt{1.02}=1.00995\ldots$, which is very close to $1+\frac{0.02}{2}=1.01$. If one requires greater precision, one can add a summand that is a multiple of x^2, and even greater precision is achieved with an x^3 term.

Here we are interested in the exponential function. For e^z one obtains the best approximations by considering two, three, or more summands from the following sum:

$$1 + z + \frac{z^2}{2!} + \frac{z^3}{3!} + \cdots .$$

(Recall that $2! = 1 \cdot 2$, $3! = 1 \cdot 2 \cdot 3$, and so on.) Since the error grows smaller as more and more summands are taken for the approximation, one writes

$$e^z = 1 + z + \frac{z^2}{2!} + \frac{z^3}{3!} + \cdots .$$

There are similar formulas for the sine and cosine functions:

$$\sin z = z - \frac{z^3}{3!} + \frac{z^5}{5!} - \frac{z^7}{7!} + \cdots ,$$

$$\cos z = 1 - \frac{z^2}{2!} + \frac{z^4}{4!} - \frac{z^6}{6!} + \cdots .$$

If one now uses the formula for e^z to express e^{ix}, where i is the imaginary unit, then we obtain, keeping in mind that $i^2 = -1$ (see Chapter 94),

$$\begin{aligned} e^{ix} &= 1 + ix + \frac{(ix)^2}{2!} + \frac{(ix)^3}{3!} + \cdots \\ &= 1 - \frac{x^2}{2!} + \frac{x^4}{4!} - \frac{x^6}{6!} + \cdots \\ &\quad + i\left(x - \frac{x^3}{3!} + \frac{x^5}{5!} - \frac{x^7}{7!} + \cdots\right) \\ &= \cos x + i \sin x. \end{aligned}$$

We now choose the particular value $x = \pi$ and recall certain values of the trigonometric functions measured in radians, namely $\cos \pi = -1$ and $\sin \pi = 0$. Thus we indeed have $e^{i\pi} = -1$, which is essentially Euler's formula.

Chapter 56

The First Really Complicated Number

There are two reasons why the numbers representable as fractions—the rational numbers—are of fundamental importance. First, there are so many of them, and they are so densely distributed among all numbers that practically every important number can be well approximated by a rational number. For example, one can determine the amount of seed necessary to plant a circular field to sufficient accuracy by approximating π by the fraction $314/100$.

And secondly, fractions are easy to work with. One can explain to a child at a fairly young age what the fraction $5/11$ represents. The Pythagoreans of ancient Greece were even of the opinion that all numbers necessary for arithmetic and geometric problems must be rational. Even though that is not the case, with that principle they were able to describe many important phenomena. For example, the Pythagorean scale arises from the fact that pleasing relationships among tones are described by simple ratios.[1]

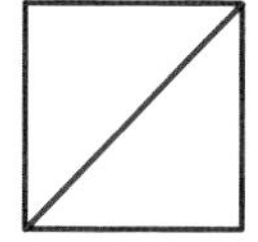

Therefore, it came as a terrible shock when it was discovered that numbers can arise in quite simple relationships that are not rational. Such numbers are called *irrational*. The best-known example is surely the square

[1] See Chapter 26.

root of 2. It arises as the length of the diagonal of a square with side of length 1. No one wishing to have anything to do with geometry can avoid it.

The proof of irrationality is not trivial. Computers and massive calculations are of no help. Just because the square root of two can't be written as a fraction with numerator and denominator of millions of digits doesn't mean that it cannot perhaps be written with billions.

The solution is to use an indirect proof. This is a method that even Sherlock Holmes frequently used to good effect: If you assume that something is true, and that leads to the conclusion that something else must be true, but this second something is actually false, then the original assumption must be false.

Such a method works in our case as well. The details appear below.

There is a story associated with the discovery of the irrationality of the square root of two. Having discovered the existence of irrational numbers, the Pythagoreans swore their members to secrecy, and the discoverer, a certain Hipposus, was put to death for having shaken the foundations of mathematics.

Why Is $\sqrt{2}$ Not Representable as a Fraction?

The square root of two is the positive number whose square is equal to 2. Let us call it r for short. A bit of experimentation can give us an idea of this number's size. For example, the square of 1.4, that is, the number $1.4 \cdot 1.4 = 1.96$, is smaller than 2, and therefore r must be greater than 1.4. On the other hand, the square of 1.5 is 2.25, which is too large. Therefore, r is certainly smaller than 1.5.

Any pocket calculator can give a more accurate value. An approximation to the actual value that would be adequate for most practical purposes is 1.414213562. That is not quite exact, since

$$1.414213562 \cdot 1.414213562 = 1.999999998944727844,$$

which is just a bit too small.

It was over two thousand years ago that mathematicians began to wonder whether the square root of two could possibly be expressed as a fraction.[2]

The following proof uses only the fact that the square of an odd number is odd, and the square of an even number is even.

The proof begins with the assumption that r can be written as a fraction, and then one makes deductions from this assumption until one arrives at nonsense. (Thus would Sherlock Holmes have argued: if the murderer had left the restaurant through the kitchen, he would have been seen by the cooks. They saw nothing. Therefore he must have escaped by another route.)

We now write r as p/q with natural numbers p and q. We cancel all common factors, and therefore at least one of p and q must be an odd number.

From $r = p/q$ we have $p = r \cdot q$, and if we square this equation, keeping in mind that $r \cdot r = r^2 = 2$, we obtain $2 \cdot q^2 = p^2$. Therefore, p^2 is an even number, and as we noted above, that is possible only if p itself is an even number. We now, since p is even, can write p as $2k$ and substitute that into $2 \cdot q^2 = p^2$. This yields, since $p^2 = 4k^2$, $2 \cdot q^2 = 4 \cdot k^2$. Multiplying both sides of this equation by $\frac{1}{2}$ gives us $q^2 = 2 \cdot k^2$. Thus q^2, and hence q, is even. But that is impossible: we assumed that we had canceled all common factors from p and q, but now it turns out that both numbers are even.

We have thus shown that r cannot be written as a fraction. Not even with astronomically large numbers. Not even in 100,000 years.

---

[2] Note that every number with a finite decimal representation can be represented as a fraction. For example, 1.41 can be written 141/100. Thus a number that cannot be represented as a fraction cannot have a finite decimal representation.

Chapter 57

P = NP: In Mathematics, Is Luck Sometimes Unnecessary?

In this chapter we are going to speak about a problem for whose solution a prize of one million dollars has been offered.

As preparation, let us consider the classification of solution procedures. Everyone knows that addition is easier than multiplication. To make that precise, consider the number of digits that occur in any problem involving numbers. In adding n-digit numbers, one needs essentially n computational steps, while for multiplication, the number is $n \cdot n$. More-complicated procedures (such as those for solving systems of equations) require $n \cdot n \cdot n$ steps. In general, we say that an algorithm takes *polynomial time* if the number of computational steps is at most a power of n, that is, n^r for some fixed number r.

Such problems are generally considered "easy" to solve, since with the aid of a computer such a problem can be handled up to a considerable order of magnitude. But there is a host of problems that appear to be fundamentally more difficult. A famous example is the *traveling salesman problem*, which asks for the shortest route connecting a number of points.[1]

And among the "difficult" problems there are those that can be solved by clever guessing—that is, with a good deal of luck. For example, factoring a large number is a difficult problem, but if one happens to guess correctly one of the factors, it is easy to demonstrate that it is indeed a factor.

No one seriously believes that such problems themselves must therefore be easy, since the kind of luck required here is at the level of being able to guess the winning lottery numbers every week for a lifetime. Scandalously, however, no one has been able to prove whether this luck is in fact necessary. Over the past decades, much effort has gone into trying to solve this problem, and since the year 2000, anyone solving it will receive a one-million-dollar prize. But before you take out paper and pencil, you should realize that some of the world's best mathematicians have searched for a proof in vain.

We should emphasize here that some of the interest in a solution is connected with the fact that the answer would have great influence over the security of current encryption systems.

What Exactly Are P and NP Problems?

For an understanding of this problem, some terms need to be clarified.

What Is a P Problem? For the addition of two three-digit numbers, one needs to carry out three elementary additions, and in general, n additions for n-digit numbers. The usual method of doing multiplication is more complicated: one must execute $n \cdot n$ simple multiplications, and then some additions. The product is calculated in at most $2n^2$ computational steps. A problem (such as "calculate a sum," "calculate a product") is said to be of type P if the amount of time for solving a problem with n-digit numbers is bounded by an expression of the form $c \cdot n^r$, where c and r are some fixed constants. The letter P indicates that these are problems that can be solved in polynomial time. For example, if a problem with n-digit input can

[1] See Chapter 32.

always be solved in fewer than $1,000 \cdot n^{20}$ elementary calculations, then the problem is of type P.

It is well known that the problems in this category are "relatively simple problems" in that they can generally be solved in a reasonable amount of time on a computer.[2]

What Is an NP Problem? There exist, however, problems that involve very complicated methods for their solution. For the existence of an optimal route in the traveling salesman problem of at most a given length,[3] if there are n cities, then there are $1 \cdot 2 \cdot 3 \cdots n$ possible routes to investigate, and that cannot be bounded by an expression of the form $c \cdot n^r$, no matter how large c and r might be.

But if one guesses at a solution and is lucky, then the problem is quickly solved: guess a solution and then verify in at most n computational steps that the route chosen is not too long.

In general, a problem is said to be of type NP (nondeterministic polynomial) if it can be reduced to a polynomial problem by guessing and luck.

It is generally considered a scandal that no one has yet been able to prove that the class of P problems and the class of NP problems are not the same. Of particular interest is the question whether the problem "find a factor of an integer" belongs to the class P. (As already mentioned, it is certainly an NP problem.) In Chapter 23 we discussed how the security of encryption systems depends on the answer to this question.

For an answer to the problem "P = NP?" the Clay Mathematics Institute has offered a one-million-dollar prize. Details can be found at http://www.claymath.org/.

---

[2]This must be considered only a rule of thumb. If the bound is $1{,}000 \cdot n^{20}$, then the solution of a 5-digit problem will require 95,367,431,640,625,000 calculations. That is more than any computer can handle.

[3]See Chapter 32. That chapter also contains some discussion on the P = NP question.

Chapter 58

Happy 32nd Birthday!

A problem can be best understood when approached from an optimal point of view. This is certainly the case in mathematics, where much effort is expended to provide a great variety of ways to represent the objects under consideration so that a satisfactory choice of method can be made when a particular problem arises.

Let us take as an example the integers. We are accustomed to writing them in the familiar base-10 system. This means that we specify a particular number by indicating how many ones, tens, hundreds, and so on one needs to represent it. Thus 405 is shorthand for "four times one hundred plus zero times ten plus five times 1."

This is a very useful form of representation, since it allows complicated calculations to be reduced to simple multiplication and addition, the teaching of which occupies a large amount of time in elementary school.

But why base 10? Certainly this comes from the fact that we have ten fingers (including the thumbs); there is no deeper reason. Thus it is that some cultures once used a system with base 12. In such a system, there are twelve symbols for the digits, let us call them $0, 1, 2, 3, 4, 5, 6, 7, 8, 9, \mathrm{A}, \mathrm{B}$, and numbers are represented using powers of twelve. We would find it strange to work with such numbers, but such a system would have certain advantages. Namely, twelve has more divisors than ten, and therefore there would be fewer situations in which a number had to be represented by a fraction.

Today, the only number systems in wide use besides base ten are those in base two and base sixteen: the binary and hexadecimal systems. Both are used in relation to computers. The binary system is practical because it uses only two digits (0 and 1), and therefore numbers can be easily represented in an electronic environment (on or off; high voltage or low voltage). And the hexadecimal system arises when one combines four binary digits into a single new digit.

For example, the number 50 is written in the hexadecimal system as 32 (namely as two times one plus three times sixteen). Thus can one transform one's 50th birthday into one's 32nd: it is simply a matter of point of view.

The Fountain in the New National Gallery

The base-3 number system has been given an artistic representation in Berlin's new National Gallery. The fountain in the inner courtyard, by the American minimalist Walter de Maria, comprises a number of small columns that come in three different forms. If these forms are interpreted as digits in the base-3 number system, then one has all possible combinations of digits, representing the numbers from zero to $3 \cdot 3 \cdot 3 - 1 = 26$ in base 3.

Figure 1. Twenty-seven numbers in base three.

How Do I Convert?

Those who would like to convert their birthday or some other number into the hexadecimal system may use the following procedure. Suppose a number is given in the usual decimal system. For example, let us use 730.

Step 1. Divide the number by 16, and consider first the remainder that results. In our example, we obtain 730 divided by 16 equals 45, with remainder 10. Since the digits in the hexadecimal system are $0, 1, 2, 3, 4, 5, 6, 7, 8, 9, \mathrm{A}, \mathrm{B}, \mathrm{C}, \mathrm{D}, \mathrm{E}, \mathrm{F}$, the digit in the ones place in the hexadecimal representation of 730 is A. It will be the rightmost digit in the new representation.

Step 2. Now we consider the quotient of the division that was done in Step 1. That number was 45. We proceed with this number exactly as we did with the original 730. We divide 45 by 16 and keep track of the quotient and remainder. The result is $45 \div 16 = 2$ with remainder 13. The remainder gives the number in the sixteens place, the second digit from the right, which will therefore be D.

The procedure continues in this way until the result of the division is a number less than sixteen. That number will be the leading digit. In our example, we were done after two divisions, since the quotient 2 is certainly less than 16. We have therefore determined that the decimal number 730 is represented in the hexadecimal system as $2\mathrm{DA}_{\mathrm{H}}$, where the subscript H reminds us that we have here a number in hexadecimal representation. Without the H, we might sometimes become confused, since, for example, in the representation of some numbers, such as our 50th birthday, which is 32_{H}, none of the special hexadecimal digits $\mathrm{A}, \mathrm{B}, \mathrm{C}, \mathrm{D}, \mathrm{E}, \mathrm{F}$ appear.

Chapter 59

Buffon's Needle

Today we are heading back 250 years in time and across the border to France. At that time and place, science had a high social standing. Many noblemen were interested in the latest advances in the burgeoning fields of science and mathematics, and a number of them made original contributions. In addition to a riding stable, any self-respecting gentleman also had a scientific laboratory, and any scientist who was passing by the estate was welcome.

One of these science enthusiasts was Georges-Louis Leclerc, Comte de Buffon, who was born in 1707 and died in 1788, one year before the start of the French Revolution. His numerous encyclopedic works, in which the knowledge of the time was gathered, are today almost forgotten. However, his name is enshrined in the history of mathematics on account of a famous experiment.

Imagine a flat surface on which equidistant parallel lines have been drawn. It could be a ruled sheet of paper lying on a table or perhaps a wide-pine-board floor. Now a needle is tossed into the air and allowed to land on the surface. Believe it or not, one can actually calculate the probability that the needle will be touching one of the parallel lines. What is more surprising is that this probability involves the number π, which is the ratio of the circumference of a circle to its diameter. This fact provides an unexpected experimental opportunity to determine the value of π experimentally. One has only to toss the needle enough times to measure the probability of hitting one of the lines with sufficient accuracy.

The method described by Buffon has entered almost every branch of mathematics under the name "Monte Carlo method."[1] The vagaries of chance are harnessed for counting, computing integrals, and much more. To be sure, no one is tossing needles around, now that we have computers, which can carry out millions of simulated random events in the twinkling of an eye.

It is a pity that today science has become so complex that those with plenty of time and money don't make use of those assets as did Monsieur le Comte.

The Formula for the Probability of Hitting a Line

The formula giving the connection between the probability of hitting a line and the number π can be derived by looking at the problem from the proper point of view.

To begin with, we need some notation. The floorboards will have width d, and the needle will have length ℓ. To ensure that at most one line is hit by the needle, we shall assume that ℓ is smaller than d. Figure 1 depicts the setup.

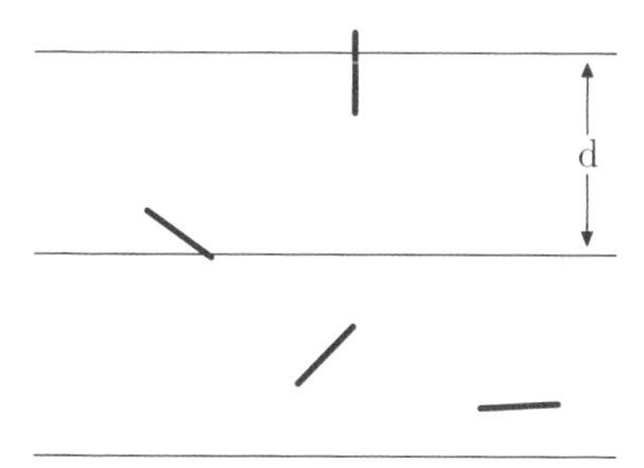

Figure 1. The setup for the needle experiment.

[1] See also Chapter 73.

Now we let chance go to work. Imagine a rectangle with sides of lengths 90 and $d/2$, drawn in the first quadrant of the Cartesian coordinate system. A point in this rectangle can then be represented by two numbers, α and y, with α between 0 and 90, and y between 0 and $d/2$.

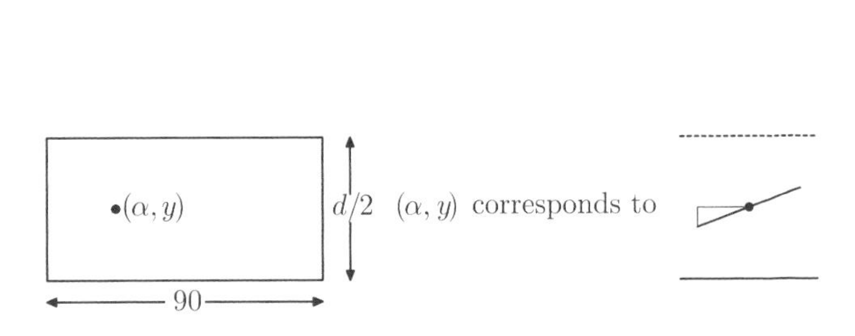

Figure 2. A random toss corresponds to a point in the rectangle.

Now the numbers α and y can be used to represent an event of the needle landing on the floor. Let y denote the distance from the midpoint of the needle to the nearest line, and let α represent the angle that the needle makes to the parallel lines. The situation is depicted in Figure 2. If α is small, the needle is almost parallel, while for $\alpha = 90$, it is perpendicular. It is then clear that for small values of y (the midpoint of the needle is near a line), a small value of α will suffice to put the needle over the line. The exact relationship can be described with elementary trigonometry: a line is hit by the needle precisely when the vertical side of the triangle shown in the figure is greater than y. But the length of this side divided by $\ell/2$ is the sine of the angle α. Therefore, a line will be crossed by the needle precisely when $\frac{\ell}{2} \cdot \sin\alpha$ is greater than y. The points (α, y) for which that happens are shown in gray in Figure 3.

Instead of actually tossing a needle, let us select points in the triangle at random and interpret them as though they were the result of a needle toss. The probability that a line will be crossed can then be read off from the figure: it is equal to the ratio of the gray region under the sine curve to the area of the entire rectangle. This area can be calculated, and we obtain that the probability of a randomly

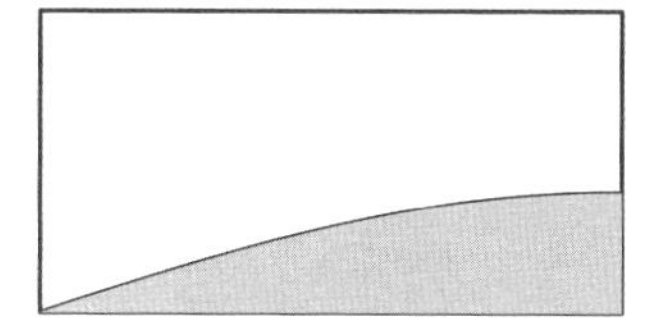

Figure 3. The points in the gray region represent "hits."

tossed needle of length ℓ crossing a line on the floor is

$$\frac{2 \cdot \ell}{\pi \cdot d},$$

where d is the width of a floorboard.[2]

Qualitatively, the formula makes sense. The probability should be greater for larger values of ℓ, and it should shrink as the width d of a floorboard increases.

So much for theory; let the π experiment begin! We toss a knitting needle (10 inches long) 1,000 times and note how often it hits a line (20-inch boards). Suppose we got 320 hits. This gives us an estimate of the probability P: it should be about $320/1{,}000 = 0.32$. And if we take the theoretical formula that we derived earlier, namely

$$P = \frac{2 \cdot 10}{20 \cdot \pi},$$

and solve for π, we obtain

$$\pi = \frac{2 \cdot 10}{20 \cdot P} \approx \frac{2 \cdot 10}{20 \cdot 0.32} = 3.125.$$

Thus the needle experiment allows us to estimate the value of π to be $\pi \approx 3.125$. Admittedly, our estimate is not very precise, but if you want a better result, simply toss the needle a larger number of times.

[2]This formula is obtained by integration. The number π appears because in radian measure, the angle 90° is equal to $\pi/2$.

Chapter 60

Running Hot and Cold: Controlled Cooling Solves Optimization Problems

Not long ago, a technical term from the outside world found its way into mathematics. The term "annealing" comes from glass manufacture. It is the process by which glass is slowly cooled to increase hardness and reduce brittleness.

In mathematics, *simulated annealing* became established as a universal tool in solving difficult optimization problems. The idea is to find a certain number of parameters that make a quantity being sought as large as possible. Say the parameters are latitude and longitude and the quantity sought is the height above sea level in a particular region. The task is then to find the highest point in the region. (Of course, it could also be the lowest point that is being sought.) The problem could as well involve the proportions of various chemicals in a reaction or the parameters of a motor: perhaps one is looking for a material with particularly good properties or the setting giving the greatest efficiency.

The classical way of solving such problems is with differential equations. But such an approach often fails because the relationship between the initial values and the goal function is not sufficiently well known or is much too complicated for concrete calculations.

In such circumstances, simulated annealing can help. This method can be explained in terms of our first example, in which a hiker is seeking the highest point in hilly terrain. Imagine that fog has set in. How are you going to find the highest point? Keep going upward? Then you might end up atop a small hill, even though much higher points were reachable. The idea of a solution is to walk upward for the most part, but every now and then, deliberately take a step downward. This gives you a chance of actually finding the highest point. You just have to make sure that having reached the highest point, you don't abandon it. This is accomplished by letting the tendency to walk downward approach zero over time. That is where the analogy to controlled cooling enters the picture.

In other problems the approach is similar. Instead of wandering across the countryside, you wander among a field of parameters. If the field isn't too large and you have adequate time for computations, you should be able to solve your optimization problem.

The Traveling Salesman

It is common to use the analogy of a journey in formulating an optimization problem. As an example, let us revisit the traveling salesman problem of Chapter 32. Let us say that we are given twenty cities, and we seek the shortest route that visits each city exactly once. We number the cities $1, 2, \ldots, 20$, and indicate a route by the order of the cities. Thus

$$6,\ 1,\ 19,\ 2,\ 15,\ 12,\ 3,\ 5,\ 20,\ 11,\ 16,\ 10,\ 7,\ 13,\ 8,\ 4,\ 9,\ 17,\ 14,\ 18$$

denotes the route that begins at city 6, then visits city 1, continues to city 19, and so on. We shall assume that we return to the initial city, so after visiting city 18, the weary salesman returns to city 6, which completes the circuit.

The number of such trips is astronomically high. With elementary combinatorics (see Chapter 29), we calculate that there are 2,432,902,008,176,640,000 possible routes. It is unrealistic to think that a computer could do the calculations to determine the length of each of these routes. To use simulated annealing, we might think of each possible route as a point in a hilly terrain, with the length of the circuit the height above sea level. We would like to find the lowest point.

With simulated annealing one would attempt to find a solution by starting with a particular circuit, say the one given above $(6, 1, 19, \dots)$. Choose two positions in the list at random, and exchange their places in the circuit. For example, if positions 3 and 8 were selected, then cities 19 and 5 would exchange places, resulting in the route

$$6,\ 1,\ 5,\ 2,\ 15,\ 12,\ 3,\ 19,\ 20,\ 11,\ 16,\ 10,\ 7,\ 13,\ 8,\ 4,\ 9,\ 17,\ 14,\ 18.$$

If the exchange leads to a shorter route, then continue the process, choosing two more positions at random. Otherwise, return to the previous route and try again. What is added to the mix, however, is this: one accepts with a certain probability a route that is longer than the previous route, and this probability is greater at the beginning of the process than at the end. This ensures that one doesn't get stuck in a local minimum (thus one really can find the lowest point, not just a depression in the middle of a plateau).

Let us consider a particular example in which simulated annealing was used. We begin with twenty "cities" in the plane, as shown in the left-hand frame of Figure 1. The distance between two cities is the straight-line distance between the points, "as the crow flies." In other problems, a different measure could be used, such as the length of the automobile route, or the cost of an airline ticket. Then a route is proposed, where we will employ a random process to do the choosing for us. The result of the random choice is shown in the right-hand frame of Figure 1.

Now we apply the simulated annealing algorithm. It searches, as was described above, through variations in the current route and as a rule, follows only those proposed alternative routes that are shorter. After a few milliseconds of computation, the computer has found the route depicted in Figure 2.

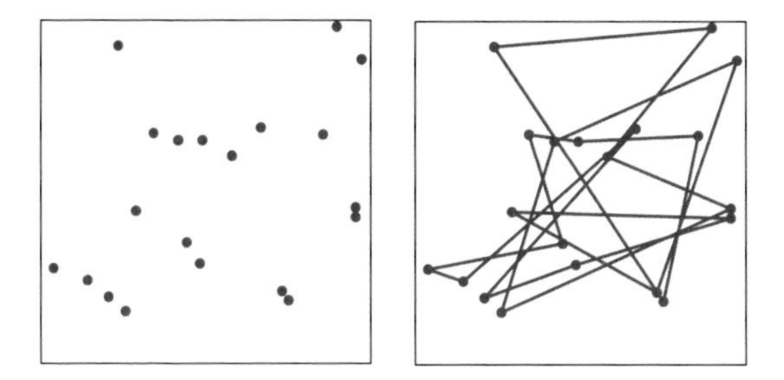

Figure 1. The cities and a randomly proposed route.

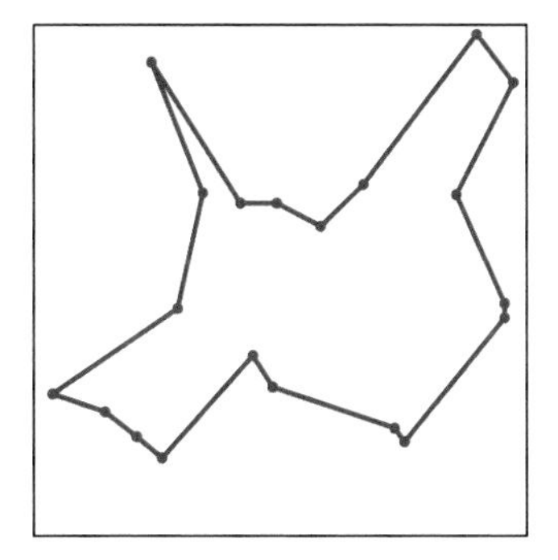

Figure 2. A route found by simulated annealing.

This looks like a promising route, and there is likely no other route that is much better.

It must be noted that one can never be certain that a better route was not missed. A guarantee of the best route would require much greater computational cost and a more complex mathematical theory.

Chapter 61

Who Didn't Pay?

It happens now and then in mathematics that one can rigorously prove that there exist objects possessing certain properties, yet one is unable to exhibit a single concrete example of such an object. In such cases, such a demonstration is called a "nonconstructive existence proof."

As an example, let us consider the classical result that there are infinitely many prime numbers. It follows that there exist prime numbers of more than 100 trillion digits.[1] However, we are far from being able to bring such giants to paper. The record holder among identified primes has "only" about ten million digits. And we may assume that the situation will not change significantly in our lifetimes.

From proof of existence to concrete examples is frequently a long and tortuous path—if indeed any examples at all can be found. For example, it was clear from the arguments put forth by Georg Cantor, the creator of set theory, that among all numbers, the vast majority are "very complicated," that is, are *transcendental*. But it took a great deal of effort before anyone was able to identify a particular transcendental number. (And it was even more difficult to demonstrate that certain well-known numbers are transcendental; in the case of the number π, the mathematician who showed that it is transcendental, Lindemann, secured for himself a place on the mathematical Olympus.) (See Chapter 48.)

[1] See Chapter 4 for more on the infinitude of primes.

One might think that comparable problems could not possibly exist in "real" life, but such is not the case. To find an example, let us step into the nearest jazz club. There are one hundred patrons swinging to the beat. At the entrance, the cashier notes that only ninety entrance fees were received. What does this mean? One may be quite certain that ten of the visitors somehow got in without paying, but do you think you could identify a single one of them?

Pigeons and Pigeonholes

There is an important method of proof that goes under the beguiling name *pigeonhole principle*. It is frequently used in mathematical nonconstructive existence proofs.

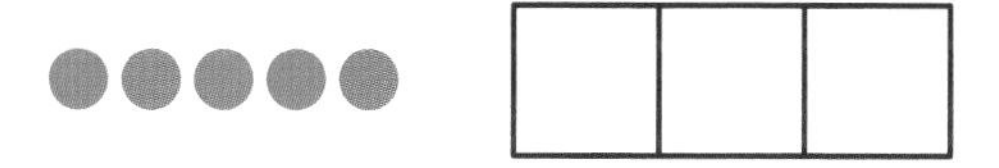

Figure 1. Five balls in three drawers: at least one drawer contains at least two balls.

The idea is simple. If a dresser has n drawers and more than n balls are to be stored in the dresser, then there must be at least one drawer containing at least two balls. One can be certain of this, without opening the drawers and without having any idea which drawer or drawers contain more than one ball.

The principle should be clear to all, even those with no experience of storing balls in dresser drawers. For example, if you have placed three handkerchiefs in your two trouser pockets, then one pocket must have at least two handkerchiefs.

> And how does one go about proving the pigeonhole principle? Remarkably, the result cannot be proved directly. It is necessary to prove what in logic is called the contrapositive, a method beloved of the detective Sherlock Holmes: If Mr. X were not the perpetrator, then Mrs. Y would have seen him. But Mrs. Y didn't see him, and so Mr. X must be guilty.
>
> In our case, we prove the following: If there is at most one ball in each drawer, then there can be only

> at most n balls. But there are more than n balls, and therefore "at most one ball in a drawer" cannot hold.

A typical mathematical application might look like this: You are given eleven random natural numbers. It is then certain that at least two of them end with the same digit. Simply imagine ten drawers labeled $0, 1, 2, \ldots, 9$ and each of the eleven numbers is placed in the drawer corresponding to its final digit.

One might also think of n pigeonholes that must accommodate more than n pigeons. Then at least one of the pigeonholes must house at least a pair of pigeons. It must have been a mathematical pigeon fancier who gave the name "pigeonhole principle" to this method of proof.

Chapter 62

What Can Statistics Tell Us?

Almost daily—even in serious newspapers—one can discover information about the latest statistical findings: researchers have discovered that mathematicians live longer than physicists; bicycle riding is more dangerous than walking; and so on.

Where do these claims come from? What is statistics capable of telling us? Alas, the answer is rather sobering, so let us look at a particular example.

Suppose you go to your favorite game store and buy yourself a new pair of dice. The sales associate assures you that each die is perfectly balanced. Arriving home, you try them out, only to discover that when you roll them, one die consistently turns up 3. Should you demand an exchange? Is the die loaded?

Figure 1. Is there something fishy about this scenario?

The question is not easily answered, since with a certain (extremely small) probability, an absolutely fair die can roll one hundred 3's in a row.[1] Now, extremely improbable events are normally not expected to occur, and therefore one proceeds as follows. Before testing the die, choose a set M of outcomes that is expected to occur with, say, 99% probability when a die is thrown one hundred times. For example, M could be the set of outcomes in which a 3 appears no more frequently than forty times. Then if your new die produces a 3 one hundred times out of one hundred (or even only 45), you may assume that you were sold a defective item. To be sure, you may be wrong, but it is very unlikely (at most 1%).

All of this is usually couched in the mathematical terminology of null hypothesis and confidence interval, but the general idea is always the same: rely on unlikely events not occurring.

Of course, the fairness of a die is a harmless enough example, but there are questions of reliability in which much more is at stake. The effectiveness of a new medicine, the probability of contracting cancer from living near a wind generator, the dangers from passive smoking, all are statistically evaluated according to the same principle.

Unfortunately, the accurate and carefully worded formulations of the statisticians are frequently lost somewhere between the journalist and the newspaper column. The reason for this is not hard to discover: what can be stated confidently and with good conscience is not as a rule particularly spectacular news.

Should I Switch Suppliers?

For a realistic example of statistical methods let us assume the role of the buyer for a factory that manufactures radios. A shipment of one thousand transistors has just arrived. Is the percentage of defective units less than 3%, as guaranteed by the supplier?

[1] Translator's note: For a literary exploration of this kind of unlikelihood, see Tom Stoppard's play *Rosencrantz and Guildenstern Are Dead.*

Of course, you could test every transistor, but first of all, that would take too much time, and secondly, the testing process often destroys the transistor. Therefore, you decide to test twenty transistors. The result: two are defective.

The buyer now uses the following reasoning process: *If* the defect rate is really at most 3%, how likely was what I have just observed to occur? He draws a table, like the one below, in which one can read the probability of $1, 2, 3, 4$ defective transistors out of twenty among one hundred items if the defect rate is 3%:

Defective Items	0	1	2	3	4
Probability	0.55	0.33	0.10	0.02	0.003

One can read off, for example, that the probability of two defective items out of twenty under the given conditions is 0.1, that is, ten percent.

That is not a particularly unlikely outcome, and so there are no reasonable grounds for rejecting the shipment. A result of four defective transistors would have been much more alarming. Of course, such an outcome is not impossible, since three percent of the transistors are permitted to be defective, and three percent of one thousand is thirty, and it could be just by chance that more than the expected number of defective pieces turned up in the test sample of twenty. But at three-tenths of one percent, the observed outcome is extremely unlikely, and therefore one can no longer simply act on the assumption of "defect rate at most 3%."

One's level of mistrust can be quantified.[2] One should be more distrustful of a new supplier with whom one has no positive history. Even with a moderately unlikely outcome one might do well not to accept the shipment. But if the supplier is one with whom you have been dealing confidently for years, the result would have to be extremely unlikely for you to doubt the guaranteed defect rate.

---

[2]The technical term is *confidence level*.

Chapter 63

Arbitrage

One of the key words in the vocabulary of the mathematics of finance is *arbitrage*. There are two things that you need to know: First, the definition. Arbitrage means the possibility of making a profit without risk and without the investment of capital. For example, if Bank A is selling dollars for 0.9 euros each and bank B is buying dollars for 1 euro, then you should quickly borrow 900 euros from somewhere. Use them to buy 1,000 dollars, sell these for 1,000 euros, and return the 900 borrowed euros. You have made a quick profit of 100 euros. It would be better, of course, if you could borrow 9,000 euros or even 90,000. Arbitrage is like the goose that laid the golden egg.

Unfortunately, the second thing you need to know about arbitrage is that there is no such thing. This fundamental law of the mathematics of finance is equivalent to the well-known maxim, "there is no free lunch." However, this law is not so strict as a basic law of physics. If, for example, the exchange rate in Hong Kong is even a little bit different from that in Frankfurt, vast sums will be moved to take advantage of the arbitrage opportunity. The percentage difference might be minimal, but with several billion euros the amount at stake can become significant. For you and me this will never be a way to get rich, since the bank fees would be greater than the profit.

And now to the mathematics. The "no free lunch" principle for finance is used like Newton's laws of motion or the second law of thermodynamics in physics. Its purpose is to figure out formulas for prices of all possible options. This form of investment has become more and more important, since options allow one to insure against too much risk. (More on this in the next chapter.) The arbitrage principle plays an important role in options trading. It is used as follows: Only when the price for a certain option assumes a certain special value does arbitrage disappear. Therefore, that is the price that should be placed on the option.

Several years ago a Nobel Prize was awarded for calculations—admittedly very complex ones—along these lines. It was given for the derivation of the Black–Scholes formula, which plays a fundamental role in options pricing.

Arbitrage as "Natural Law"

The principle that there is no such thing as arbitrage plays the same role in the mathematics of finance as a natural law (for example, force equals mass times acceleration) in physics. One can use it to make new discoveries.

As an example, let us consider an arrangement whereby during the coming year I receive a guaranteed payment of 100,000 euros. The payment could be related, for example, to the transfer of a complex stock portfolio that has been guaranteed to be priced at 100,000 euros at the end of the contract period. How much should I pay for such a contract?

Suppose that I can borrow money at 4% interest.[1] The arbitrage principle implies, then, that the contract has precisely the value $100{,}000/1.04 = 96{,}154$ euros. Here is why:

- What if the contract were available for less, say 90,000 euros? Then I would borrow 90,000 euros from the bank and purchase the contract. After a year, I would get my 100,000 euros. I would immediately pay my debt to the bank: 90,000 euros principal plus interest of $90{,}000 \cdot 0.04$ for a total of 93,600. There is 6,700 euros left for me, a completely risk-free profit. Arbitrage! But since arbitrage doesn't exist, no

[1] And we shall assume as well that I can get 4% interest on savings.

one is offering such a contract for 90,000 dollars. And the same argument holds for any price less than 96,154 euros.

- What would happen if such contracts were available at prices higher than 96,154 euros, say for 98,000? Then I would begin selling such contracts at that price. The customer pays me 98,000 euros. I deposit 96,154 in the bank and squander the remaining $98{,}000 - 96{,}154 = 1{,}846$ euros. Arbitrage! I can easily satisfy my obligations, since at the end of the contract term, my 96,154 has grown to a value of 100,000, which I pay to my client.

  The moral is this: Arbitrage exists at prices above 96,154 euros, and therefore such prices cannot occur.

Thus there is a single price, 96,154 euros, that does not lead to arbitrage, and that is therefore the only fair value that can be placed on such a contract.

Chapter 64

Farewell to Risk: Options

Suppose you are the owner of a vineyard that reliably produces about ten tons of grapes every year. They are bought by a vintner, since you know plenty about growing grapes but nothing about making wine.

Unfortunately, you cannot be certain how much your grapes will bring at market come harvest time. To ensure a reasonable return, you would like to have some sort of "insurance policy." You come up with what seems like a reasonable selling price, P, for your grapes, and then you set about looking for someone with whom you can make the following contract: You pay your partner a certain sum for signing the contract. If in the fall, the price of grapes is less than P, the partner pays you the difference. If the price is higher than P, you make the profit and your partner owes you nothing.

Such arrangements are made daily by the tens of thousands; they are called *options*. They are contracts that attempt to reduce or eliminate the risk of an uncertain outcome. Almost anything can be thus insured: market price of grapes, sugar cane, and gold; selling price of the dollar, electricity, telecom shares; and so on. Options have become a matter-of-fact commodity. For example, you can go to your bank and purchase an option to buy ten thousand shares of a telecom stock on 3 October at 20 euros a share. If the listed price on that date is lower, the bank is pleased, since it knows that you are not going to buy the shares. If the listed price is higher, the bank pays you the difference. And no one will ask whether you actually buy the stock or use the money for a vacation.

Mathematics comes into play because the partner in such a contract has to know how much the contract is worth to you. In making such a calculation, one of course keeps in mind the law of no arbitrage, as was discussed in the previous chapter. Namely, no one should be able to make a risk-free profit. After input of the relevant parameters—interest rates, expected fluctuations in the stock market, target sales price, term of the contract—the price can be easily computed.

Since there is an enormous variety of options available, with new ones appearing daily, mathematicians have plenty to do. Large banks employ them by the hundred, and research is being done at universities on creating better models that make more precise predictions.

And now a warning: Options trading is very seductive, since with some luck, one can double one's investment in a few weeks. But sometimes, it goes the other way, and your money is gone. Therefore, weekend gamblers might want to stick to betting a few euros on the national lottery.

Put or Call?

The international language of finance is—one could almost say "but of course"—English. Here are some of the most common terms that one is likely to encounter.

If one is interested in selling something, then *put options* are of interest. This is the case, for example, with our vineyard owner described above. One then has to come to terms with the bank over certain details: How much should be paid for the grapes at the end of the contract term (which comes into play, of course, only if the going price is lower than that set in the contract)? That is the *strike price*. As is logical, a higher strike price makes the option more expensive.

The most important varieties of "payment rules" are *European* and *American* put options. With European options, there is a fixed date on which the contract terminates. For example, the payout amount could depend on the selling price of grapes on 31 October. With American options, in contrast, one can go to the bank at any time up to a certain date, say the end of July, and redeem the contract. One would do so if the price of grapes were particularly low.

Those interested in buying something will find *call options* just the thing. If on 13 December I will need five tons of sugar cubes, I can guarantee the price with a call option, say for a price of 2,000 euros. If it should happen that in December the global price of sugar has risen and I would have had to pay 2,500 euros, the bank will have to pay me the 500-euro difference. In this case as well there are both European and American variants, and again it is clear that the lower the strike price, the more expensive the option.

By the way, one can get involved in options even if one has no use for say, five tons of sugar cubes but would like to get rich through pure speculation. In fact, an ever decreasing proportion of options are related to some underlying commodity.

Chapter 65

Is Mathematics a Reflection of the World?

Do we practice the "correct" mathematics? The naive answer is a resounding yes, since many mathematical laws are modeled after our experience of the real world, and the world reflects our mathematical laws. For example, the abstract result "inequalities can be added" is in harmony with our experience that buying your groceries at the supermarket is cheaper than shopping at Balducci's. Since each article is cheaper in the supermarket, the total at the register must also be less.

The answer is no longer so clear when once passes from simple numbers to more complicated objects. For example, one has to know the precise fundamental laws that define functions to be able to demonstrate rigorously that a continuous function must take on the value zero at some point if it has both positive and negative values. That this must be so is "clear" to everyone, but mathematicians are not content until they have found an airtight proof. It is even more complicated to show that in Figure 1, any path from A to B must intersect the circle.

Indeed, as everyone "knows," there must be a point at which the path crosses the circle. The correct conceptualization of the problem with an exact formulation and proof was accomplished a mere 150

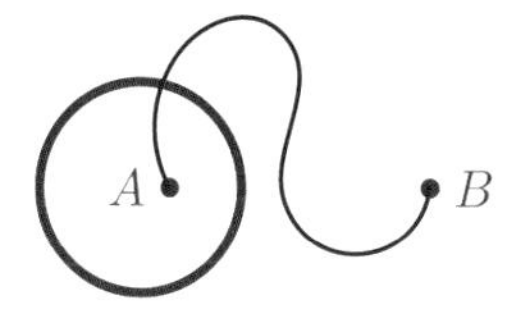

Figure 1. Every path from A to B must cross the circle.

years ago. The problem consists of two parts: First of all, what exactly is a path connecting two points, and how does one express the fact that there are no "breaks" in the path? And second, once that has been made precise, how does one then prove the existence of points of intersection?

Sometimes, a very long time passes between the formulation of a problem and a satisfactory solution. A famous example comes from knot theory.[1] Again, everyone "knows" that there are knots that cannot be untied, no matter how clever one may be. Yet it took a great deal of effort before this obvious fact of everyday experience was elevated to a mathematical theorem.

The formulation and proof of "obvious" facts as mathematical theorems is necessary because experience and intuition cannot be trusted completely. Things become even more complicated when we enter domains to which we have no direct sensory contact. Consider, for example, the realm of the infinite. There we discover laws that astound us. For example, in a way that can be defined rigorously, an edge of a rectangle has the same number of points as the entire interior of the rectangle.

Moreover, in describing what goes on at cosmic or microscopic distances according to the current state of science, mathematical models are necessary that are quite incomprehensible to the layperson. It is only with such models that one can grasp the four-dimensional spacetime of the theory of general relativity or the laws of quantum mechanics.

---

[1] More on knot theory can be found in Chapter 76.

In this sense, mathematics makes the "correct" building blocks available, but which of these should be used for modeling the world often becomes apparent only after a long search.

The Doubling of the Orange

Even in the mathematical description of phenomena that are not accessible to our direct sensory perception, there are consequences that agree with the expectations of normal human intelligence. However, sometimes one is forced to the realization that there are results that were not to be expected from a naive application of everyday experience. The concept of equality of infinite sets in general use today tells us that the number of elements of an infinite set is not reduced if I remove three, or even three thousand, elements (cf. Chapter 78).

And it can get even more dramatic. In the example about infinity, we can console ourselves with the thought that such paradoxes relate to a realm for which our genes were evolutionarily unprepared. Yet there are paradoxes as well that deal with quite elementary concepts. A famous example is the *Banach–Tarski paradox*. It states that using generally accepted methods, one can divide a sphere—for example, an orange—in such a way that the pieces can be cleverly reassembled to form a sphere of twice the size. See Figure 2.

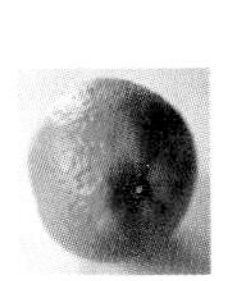

Figure 2. Is it magic?

It takes a measure of precise analysis to see that the truth of the assertion only seems to be impossible. The ability to double the orange has to do with the fact that in "cutting" the sphere, pieces result that are so full of zigzags that there is no way of associating with them a reasonable notion of volume. And therefore one cannot make use of the argument that no matter how the pieces are reassembled, the volume must remain unchanged.

If it should happen in the future that the methods of mathematics yield results that are "incompatible" with the world, then for better or worse, some necessary reconstruction of the foundations will be necessary.

Chapter 66

Mathematics That You Can Hear

Our hero today is Joseph Fourier, who developed what is called *Fourier analysis* at the beginning of the nineteenth century. He led a life rich in event, conditioned by the turmoil of the French Revolution. Among other things, he was with Napoleon in Egypt and was the first to write a systematic scientific report on Egyptian history and culture.

Today, Fourier analysis is a basic tool of all mathematicians and engineers. The idea behind it is the simple representation of oscillatory phenomena. We will focus here on musical tones, that is, on audible vibrations. The "atoms" of a tone are sine waves of various frequencies (see Figure 1). If you wish, you can hear such a tone right now. Just give a little whistle, and you are hearing what is almost a pure sine wave.

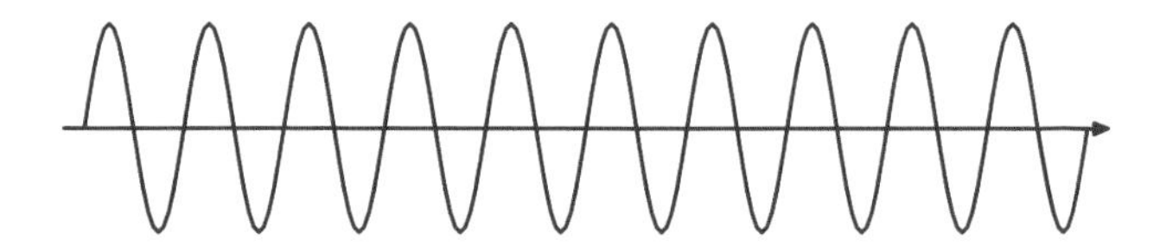

Figure 1. A sine wave.

The theory now tells us the various intensities of different sine waves that must be added together to obtain a prescribed waveform. For a musical tone, one begins with the sine wave of the fundamental frequency, adds a bit of the sine wave for the doubled frequency, perhaps a bit of the threefold frequency, and so on.

And this can be verified with our sense of hearing. A waveform that consists of a sine frequency and a portion of the threefold frequency is a good approximation to the so-called square wave. In order to hear the difference between a sine wave and the square wave, the threefold frequency needs to be within the audible range, which for most readers will be at about 15 kilohertz. Therefore, the difference between the two types of waveforms should be noticeable up to a fundamental frequency of five kilohertz.

In order to verify this, one needs ideally a frequency generator (perhaps you have a friend who is an engineer). Or do you have a synthesizer or some other electronic musical instrument? Then simply choose the waveforms "sine" and "square," and the experiment can begin.

Those who must be satisfied with a qualitative verification of Fourier's theory might wish to take note of the voices the next time they are at a party. It is easier to distinguish the deep voices of men from one another than the higher female voices. That is because men's voices have a large number of overtones in the audible range, which gives the ear many chances to make a differentiation.

A Black Box

There are other mathematical results that you can verify with your ears at least qualitatively. Imagine a black box in which one can input signals, which are then somehow processed in the box's internal mechanism and then output. Electronics hobbyists can imagine some wildly complex circuit into which an electric signal is introduced at some point and measured at some other point (see Figure 2).

This black box should have the following properties:

- It should be "linear." That is, if the strength of the input signal is doubled, the output should be doubled as well, and

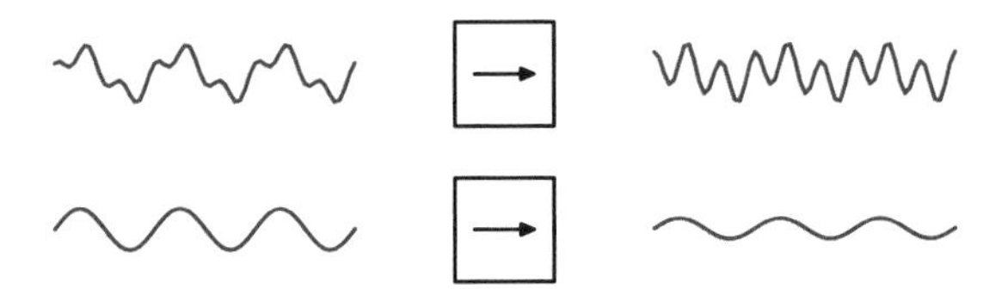

Figure 2. The input and output of a "black box."

if one inputs a signal that is the resultant of two partial signals, then the output is the same as what would result from the combination of the outputs of the two partial signals.

- It should be "time invariant." That is, if a wave is input and the output is logged, the output for a given input should be the same today as it was yesterday.

For the electronics hobbyist this means that transistors may not be used (they are not linear), and no settings can be changed during the experiment. One should limit oneself to resistors, inductors, and capacitors, and the currents and voltages that arise should not be too great.

Although such black boxes describe a rather general situation, they all have one property in common: Sine waves, the building blocks of Fourier analysis, pass through such a black box essentially unchanged. They can be weakened or pushed out of phase, but that is all that can be done with them.

The audible consequence is this: a filter for acoustic signals (high-pass, low-pass, band-pass, and so on) that can be described as a black box with the properties described above does not change the character of sine waves. If you whistle into such a filter (which gives a good approximation to a sine tone), a whistle of the same frequency should come out the other end. On the other hand, a sung tone can have its character changed completely; for example, it could be much duller or much shriller.

A Recipe for Periodic Waves: Fourier's Formula

Periodic oscillations are combined according to Fourier's theory in the form of sine waves. What is the exact recipe? That is, in what proportions do the various sine functions appear?

Suppose we have a function f, whose graph is shown in Figure 3.

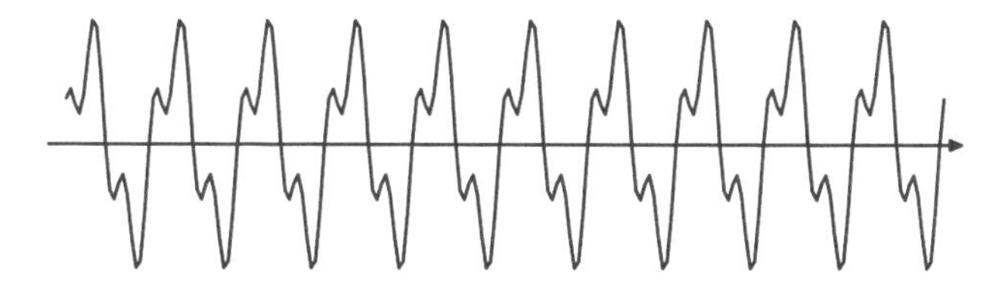

Figure 3. A periodic function f.

There is a number p (the period) such that the function evaluated at $x+p$ is always the same as at the point x. Therefore, it suffices to know the values of the function on an interval I of length p, such as the section shown in Figure 4.

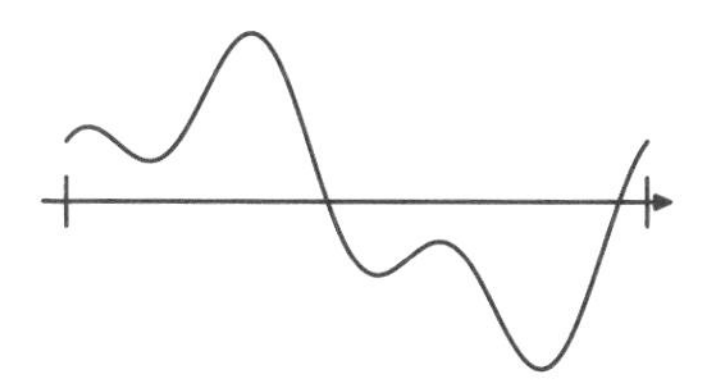

Figure 4. The significant portion of f.

One usually normalizes the function and assumes that the period is given by $p = 2 \cdot \pi$, which makes the formulas especially simple. This can be easily achieved through a change in the unit length on the x-axis.

As a final preparation, one needs to know what is meant by an *integral.* The idea is simple: If g is a function defined on an interval, then the integral of g over that interval represents the area between the graph of the function and the x-axis. Warning: any portion of the graph that lies below the x-axis is considered to have negative area. For example, if the area between the positive portion of the function and the x-axis is 4, and that between the negative values and the x-axis is 3, then the value of the integral is $4 - 3 = 1$. And if both parts are of the same size, then the integral is equal to zero. (The graph of Figure 4 shows just such a function.)

Now the "ingredients" can be calculated: If f is a function with period $2 \cdot \pi$, then one can write f as

$$f(x) = a_0 + a_1 \cos x + a_2 \cos(2x) + a_3 \cos(3x) \\ + \cdots + b_1 \sin x + b_2 \sin(2x) + b_3 \sin(3x) + \cdots,$$

where "sin" and "cos" denote the sine and cosine functions.[1] The "weights" $a_0, a_1, \ldots, b_1, b_2, \ldots$ that are used to build up the function f are determined as follows:

- a_0 is the integral of f (on the interval from 0 to 2π) divided by 2π.
- a_1 is the integral of the function $f(x) \cos x$ (on the interval from 0 to 2π) divided by π.
- a_2 is the integral of the function $f(x) \cos(2x)$ (on the interval from 0 to 2π) divided by π.
- And so on.
- b_1 is the integral of the function $f(x) \sin(2x)$ (on the interval from 0 to 2π) divided by π.
- And so on.

In sum, if you can calculate integrals, you can determine the amounts of each of the individual components that make up the periodic function.

[1] The cosine function is nothing more than a time-delayed sine function. Therefore, in the formula one could as well have only sine functions.

Chapter 67

Chance as Composer

The theme "chance as author" was considered in Chapter 10: a monkey at a typewriter will produce, given enough time, all the works of world literature.

In music, chance is used much more seriously. From Mozart (shown in the picture) we have a "dice composition" that is to be assembled according to the following rules: Roll two dice and count the total number of dots. Then choose from the collection marked "first measure" the appropriately numbered measure from among the eleven examples numbered 2 through 12. Do the same for measure 2, measure 3, and so on, until you have selected sixteen measures.

You then have only to lay the measures end to end and play the result. The finished piece certainly does not sound like the product of genius, but it might frequently be taken for part of a sonatina by one of Mozart's contemporaries.

Since for each of the sixteen measures there are eleven possible selections, there are 176 measures altogether, which can be combined into a piece according to the rules in 11^{16} different ways. However, some of the 176 measures are the same as some of the others, since Mozart sometimes used the same building block more than once.

Nonetheless, there remains the immense number 759,499,669,166,482 of compositions. Therefore, after one has finished throwing the dice, one may be fairly certain that the work created has never before been heard.

Chance plays a much more important role in contemporary music. In the music of Xenakis, for example, chance determines not only which notes are played and in what sequence, but also the waveform that is used to produce the sound.

Perhaps Xenakis's music evokes great enthusiasm in a relatively small number of listeners. However, it is interesting to ask what role chance plays in classical music. What moved Schubert, in measure six of a C-major waltz, to modulate suddenly to E major? Why did Mozart decide to set the "alla turca" finale of his A-major sonata in A minor? Is it a question of a genius who takes inspired dictation from another world, or did perhaps the random firings of certain neurons in the brain play a role?

We are not yet able to see so deeply into the brain. But surprises are not out of the question, for in the last few decades we have achieved the insight that random influences can have quite a productive and stabilizing influence.

Mozart from a Computer?

Anyone who takes the trouble to play through a large number of Mozart's dice compositions will realize that after a while, one has the feeling that one has heard all of this before, even if the notes have never been played in this particular arrangement. The reason is that our brains are capable of recognizing musical structures: What harmonies were used, and in what order? What is the rhythmic structure? What intervals are preferred? If these aspects are the same in two musical compositions, they appear very similar to us.

One can make use of this fact to program a computer, after a careful analysis of their works, to compose music that sounds like Mozart or Bach. One has simply to filter out the important aspects of the musical structure and with these parameters create something new. With what probability, say, in a C-major piece does the note

B appear after the sequence G followed by C? With what probability was it an E? Then the computer will do the same: if a G and a C have appeared in sequence, then a B or an E will follow with the prescribed probability.

To the average listener, the result sounds "something like Mozart" or "something like Bach." To be sure, there are no new ideas, and one wouldn't go so far as to call the music inspired.

Building on this approach, the composer Orm Finnendahl developed a method of creating hybrid compositions. One begins with the analysis of two compositions, A and B,[1] and initiates the hybrid composition using the parameters of A. That is, all the harmonies, rhythms, and note patterns are based on those of composition A. Then the parameters gradually begin to change, until toward the end of the piece they are those of B. The result is a piece that begins in the style of A and ends in that of B.

759,499,667,166,482 Possibilities?

For the final measure of the first part of Mozart's composition, that is, for measure 8, there are eleven different numbers from which to choose. However, the measures are all identical. Thus measure 8 is fixed in advance. Similarly, for the very last measure, number 16, there are again eleven different numbers, but only two different actual measures. For the remaining fourteen measures numbered $1, 2, 3, 4, 5, 6, 7, 9, 10, 11, 12, 13, 14, 15$, there are a full eleven different measures, and since, as we mentioned, there is one measure with two possibilities, altogether there are

$$11^{14} \cdot 2 = 759{,}499{,}667{,}166{,}482$$

different compositions.

We should note that the various compositions have different probabilities of being realized. That is because when you roll a pair of dice, the sums 2 and 12 are rather unlikely: each appears with probability only $1/36$. The middle values are seen much more often, with the

[1] Finnendahl experimented with works of the composers Josquin and Gesualdo.

probability of rolling a 7 topping the list at 1/6. Therefore, compositions all of whose measure numbers correspond to very large or very small dice rolls are extremely improbable.

Chapter 68

Do Dice Have a Guilty Conscience?

Probability can be truly confusing. Imagine rolling a single die repeatedly. On the one hand, one often hears that after "many" throws, all the different numbers come up about the same number of times on average. However, it is also said that chance has no memory, that the probabilities at any throw are the same as they were at the outset.

But how can that be? If one has rolled the die a large number of times and no six has yet appeared, wouldn't the die have to make a bit of an effort in order to produce sixes more frequently to satisfy the first condition and average things out? And so shouldn't the odds of rolling a six increase? This is the same philosophy that leads some to choose lottery numbers that have not been drawn in a long time.

This contradiction is resolved by the fact that the "equal opportunity" of all numbers does not necessarily result in equal performance: what is expected is *approximately* equal numbers with *very high probability*. One can calculate that in such an experiment the odds are close to one hundred percent that all six numbers will appear approximately equally often. However it is possible, though with vanishingly small probability, that the improbable will occur and only threes will be rolled.

To clarify what is at stake here, let us imagine an enormous number of parallel universes in each of which a dice experiment is carried out in which a die is rolled six hundred times. In the majority of the worlds, one will see that things are as they should be: each of the numbers 1 through 6 will appear very close to 100 times each. But there will be a few worlds in which the result is bizarre, where, for example, only threes appear (the proportion of these worlds is 0.000...012, with 466 zeros in front of the "12"). Or worlds in which not a single six appears (proportion is 0.00...31, with 47 zeros before the "31").

The moral is that dice are incorruptible; they have no memory. If the results are crazy, it is just that you happen to have walked in on an experiment in a most unusual world.

Our Deficient Understanding of Chance

The way in which we fail to understand the nature of chance as described in this chapter is very widespread. For children playing pachisi it is quite plausible that rolling a six is particularly likely if a six hasn't been rolled in a while. One should also mention the belief that California is due for a major earthquake, since it has been more than the average length of time since the last such catastrophe.

A small experiment should convince you of our mildly deficient understanding of chance. Take a piece of paper and write down, without tossing a coin or anything of the sort, a made-up random sequence of hypothetical coin tosses: 0 for heads, 1 for tails. Perhaps you will end up with something like the sequence below:

10011100101101000111010100101100101000111001011000101000....

A true random-number generator will produce sequences that look more like this one:

11010100111010001111011111111001111010010110110100110 00....

Do you see the difference? In a truly random sequence, it is not so improbable to see a large number of heads or tails in a row. When one tries to invent a random sequence, one can't help but unconsciously influence the outcome.

Chapter 69

Strawberry Ice Cream Can Kill You!

It's like with all expert reports: whatever it is you're trying to prove, you can always find some statistic to back you up.[1] Here are some examples.

Suppose you would like to formulate a questionnaire in which you ask whether Whit Monday should no longer be a holiday and instead revert to being a regular workday. Depending on how the question is framed, you may expect a different set of answers. From the point of view of the unions, the question would be formulated differently from its formulation from the point of view of the Chamber of Commerce: standard of living versus attractiveness of Germany to foreign investment. Both sides have a valid point of view, but in formulating the question, decisions must be made that will never be able to be corrected by mathematics.

Or think of the branch director feeling nervous about her upcoming presentation at global headquarters. Last year, sales increased only from 100,000 euros to 101,000, that is, a measly one percent. If she presents the left-hand diagram of Figure 1, it's going to look like stagnation.

[1] Translator's note: The English-language reader may be familiar with the following observation, attributed by Mark Twain in his autobiography to Benjamin Disraeli: "There are three kinds of lies: lies, damned lies, and statistics."

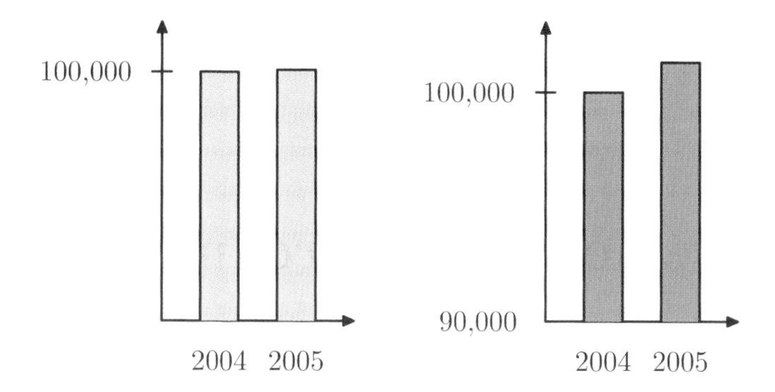

Figure 1. It's all a matter of presentation.

The solution? She should show only the top part of the bars, as in the right-hand picture. The bar from the previous year goes from 90,000 to 100,000, while this year's goes from 90,000 to 101,000. The second bar is ten percent longer than the first one, which makes things look much better.

Another way of fooling people is to cherry-pick a complex set of statistics for the favorable pieces. If a scientific study were to show that eating too much strawberry ice cream stabilizes blood pressure and dangerously raises the level of blood sugar, the editor writing the headline on the report can choose between "Better Health through Strawberry Ice Cream" and "Strawberry Ice Cream Can Make You Sick."

The moral here is that the path to the discovery and representation of truth is sown with land mines. It begins with the problem of finding a generally accepted definition of "truth."

Then the path to truth is besieged by those out for their own interests. If it finally comes down to a question of mathematical statistics, then a watertight and therefore very cautious answer can be formulated, which will then be interpreted by each according to his desires: "Strawberry Ice Cream Can Kill You."

Rich or Poor

One truly needs to defend statistics against those who see it as nothing more than a convenience store where you shop for the results that support your interests. Here are a few more examples.

It's All a Question of the Definition. Who is poor? From reading newspaper reports over the past several years, one could easily form the opinion that human misery in Germany is on the rise. Any visitor to Germany ignorant of the true situation would expect to encounter masses of starving citizens clothed in rags.

In fact, however, a large part of the phenomenon exists only on paper.[2] The reason is the definition of "poor," which is to have an income less than the average among the population. But that is a most remarkable definition of poverty, which permits at most a coarse measure of "felt poverty." It is perhaps true that young people feel left out if they don't have the latest blue jeans and cell phone, but is that really poverty that calls for strenuous action?

Random Scattering. We have spoken frequently in these chapters about the vagaries of chance. And so just as the roll of the dice can produce five sixes in a row, so it is that events that occur independently in different places can accumulate in unexpected ways.

Suppose some rare disease occurs in the German population on the order of a couple of thousand cases a year. If for each occurrence one were to stick a needle in a map at the location where the disease struck, the result would be a pattern something like that produced by a random-number generator. It is thus not at all surprising that here and there a clustering of pins will occur, and it is anything but remarkable that such a cluster might appear in the neighborhood of a highway or an atomic power plant or a city dump. But that is then taken as proof by the opponents of highways or nuclear energy, and all arguments about statistical relevance fall on deaf ears.

[2] To head off angry letters to the publisher, let it be stated here that in Germany, as in other countries, poverty unquestionably exists.

Chapter 70

Prosperity for All

We have mentioned previously that the great Galileo observed that the infinite is full of surprises. Today, by way of a thought experiment, we are going to send some chain letters in an infinite world.

The idea of a chain letter is attractive, since it promises something for (almost) nothing:[1] I send one euro to a given address and keep the chain going by sending copies of the letter to ten acquaintances. They in turn send it to ten of their acquaintances, and each of them writes ten letters. That is already one thousand people who are supposed to send me a euro. It is a pity that this marvelous scheme breaks down when there are no new acquaintances to write to.

But in an infinite world, things are different. As preparation, let us number our circle of acquaintance $1, 2, 3, 4, \ldots$. A person corresponds to each number, and the numbers can be arbitrarily large.

The game begins. Person number 1 sends the chain letter to the next ten, that is, to persons numbered 2 to 11. Each of these ten writes to ten additional persons, incorporating numbers 12 to 111. Now each of these hundred persons has to send ten letters, and the recipients will be numbers 112 to 1,111. And so it goes.

Somewhere in the chain letter is written, "send one euro to the person whose name appears three "generations" before you in the

[1] We spoke about chain letters earlier, in Chapter 6.

chain. Thus person number 1 receives one thousand letters with a euro inside from numbers 112 to 1,111. Numbers 2 to 11 receive one thousand euros each, and when it is all over, everyone in the world is at least 999 euros richer, having received 1,000 euros and sent at most one euro (from number 112 on).

Of course, no one can prevent us from playing the game not with one euro, but with ten or one hundred, or even more. Before you know it, we have all become millionaires. Is this all on the up and up? Yes indeed. Basically, this is nothing more than a shift of assets from those with higher numbers to those with lower. But since the numbers become arbitrarily high, each individual should be satisfied. What a pity that the world in which we live is finite. From a financial point of view, an infinite world would be paradise indeed.

Both a Borrower and a Lender Be

As a variant of the chain letter idea, let us consider another scheme for self-enrichment. We again number the citizenry with the integers $1, 2, 3, \ldots$. Person number 1 needs one thousand euros, so he is going to borrow this amount from person number 2. Unfortunately, person 2 is broke, but she borrows two thousand euros from person 3, lending one thousand to person 1 and keeping one thousand for herself. Alas, person 3 is low on funds. In order to supply the two thousand euros, he borrows three thousand from person 4, keeping one thousand for self-fulfillment and passing along the remaining two thousand to person 2. And so on. If there were a last person involved, he or she would end up with a mountain of debt. But in an infinite world, everyone is happy and content, and the economy just hums along.

If we replace "person" by "generation" and pass from thousands of euros to billions, we will have a fairly accurate depiction of the politics of finance in Germany (and other industrialized nations) in recent decades. The national debt is growing by orders of magnitude, and the cohesion of society in future generations depends on the assumption of further debt, whose repayment is then pushed ever further into the future.

A blemish in this scheme is that borrowing doesn't come for free. The interest on the debt also has to be financed through even more borrowing. This leads to an exponential growth in debt, and one must wonder how long the financial markets will remain willing to lend ever more money to keep the system going.

These techniques are frequently used by ordinary swindlers as well as governments. They promise fabulous interest rates and take the first million from as many trusting innocents as they can. Then they live high on the hog, keeping only enough to pay back at the end of the year the fabulous 20% interest. Word gets around that this is a great deal, and so the next millions flow in, which are enough to pay the next round of interest and a life even higher on the hog. And when the day comes that one of the first customers wants to withdraw his investment—though why would he want to do that?—there is enough money for that, too. And so it can proceed for a long time, until one fine day, the whole scheme comes crashing down.

Chapter 71

No Risk, Thank You!

Suppose you are a bank director. A customer enters your office and states that she would like to contract with you to purchase five hundred shares of a certain telecom company next 1 January. She says that she hopes and expects that the share price then will be at most 20 euros. If the price is greater, she would like you—the bank—to make up the difference.

These days, such arrangements are nothing unusual. They are called *options*.[1] For the customer, the security that the proposed contact offers is not to be had for nothing. When the contract is signed, she will pay a certain premium. What should you do with this money in order to be able to fulfill your contractual obligations come 1 January?

The magic word that allows this problem to be solved is *hedging*. In the mathematics of finance, hedging your bets amounts to the clever insuring against risk.

From the dictionary:

> **hedge:** 1. A row of closely planted shrubs or low-growing trees forming a fence or boundary. 2. A line of people or objects forming a barrier: a hedge of spectators along the sidewalk. 3. a. A means of protection or defense, especially against financial loss: a

[1] Cf. Chapter 64.

hedge against inflation. b. A securities transaction that reduces the risk on an existing investment position. 4. An intentionally noncommittal or ambiguous statement.
hedgehog: 1. Any of several small insectivorous mammals of the family Erinaceidae of Europe, Africa, and Asia, having the back covered with dense, erectile spines and characteristically rolling into a ball for protection.

The underlying idea is simple and quite clever. In addition to the money received from your customer, you borrow money at market rates and use it—customer's money plus borrowed money—to purchase telecom shares.

And to what purpose? If the share price rises by 1 January, they will be worth enough for you to be able to pay what you owe the options holder together with the money that you borrowed plus interest. If the price falls, that is a pity, but then the options holder has no demand against you, and the proceeds from the sale of the stock should suffice to pay back what you borrowed.

In sum, to insure against loss in a telecom stock transaction, some of the stock is purchased. No matter how things develop, you have placed a bet on both a rise and a fall in the stock price.

Mathematics comes into play in the determination of a fair price for the option, and how much stock the bank should purchase. Based on the "natural law of financial markets" (see Chapter 63), namely, that there can be no profit without risk, the amount in question comes from solving a simple equation. What makes things complicated is that one must keep track of the markets throughout the option period. As stock prices and interest rates change, one must determine whether some of the stock should be sold or more money borrowed for an additional purchase.

A Hedge for One Thousand Shares

Let us look at hedging through a concrete example. It is January. You would like to take out an option for the purchase of one thousand shares of Intergalactic Enterprises at the end of the year. If purchased today, the cost of the shares would be 10,000 euros. However, the price at the end of the year is uncertain. It could be 16,000 euros, but it could just as well be only 8,000, depending on a host of factors. (For the sake of simplicity, let us assume that one of these prices will obtain at year's end and that we will do no trading during the year.) Come December, you will have 12,000 euros available, and if the price is 8,000 euros, all will be well. However, if the price is 16,000 euros, you would like the bank to make up the difference. How much should the bank ask you to pay for such an insurance policy, and what should it do with the money that you pay?

The bank officer from whom you are seeking to purchase the option calls down to the credit division and learns that the bank's internal interest rate is 6%: for a distribution of $E/1.06$ euros today, one has to pay back E euros at year's end. With this information, the options contract can be drawn up. The bank requires you to pay, in return for their guarantee,

$$5{,}000 - \frac{4{,}000}{1.06} \approx 1{,}226$$

euros.[2]

You sign on the dotted line, and this is how things then develop. The credit division immediately turns over $4{,}000/1.06 \approx 3{,}774$ euros to the bank officer, who now has available $1{,}226 + 3{,}774 = 5{,}000$ euros. She uses this money to purchase 500 shares of Intergalactic and then forgets about the whole matter until December.

Suppose that the shares have risen. The bank's portfolio for this transaction is now worth 8,000 euros (recall our assumption that if the stock rises, one thousand shares would be worth 16,000 euros; the bank purchased only 500). The bank pays you the contracted 4,000 euros, and with the 12,000 euros that you had budgeted, you can now purchase 1,000 shares of Intergalactic for 16,000 euros. With the remaining 4,000 euros, the bank officer repays the credit division.

If the share price has fallen, then the bank's portfolio is worth only 4,000 euros, which is just enough to settle with the credit division.

[2]This number does not include the bank's fees, which are where its profit is realized. We are ignoring this issue.

You, the option holder, get nothing, since your 12,000 euros is more than enough to purchase the shares.

The moral: with a hedging strategy, you were able to insure a risk of 4,000 euros relatively cheaply (1,226 euros), since 4,000 is the amount you would have been short if the stock price had risen.

Chapter 72

A Nobel Prize in Mathematics?

Is there a Nobel Prize in mathematics? Until a few years ago, the answer would have been a clear and resounding no. Instead, mathematicians have their prestigious Fields Medals, which are awarded every four years at the International Congress of Mathematicians. Even though the recipients of these awards can be assured of financial stability, since they are certain to be showered with offers of well-paid positions, the actual monetary value of the prize itself is quite modest. The prize for the best young poet in the city of Wanne-Eickel is worth more.

But all that has changed in the last few years, though the prehistory of the prize goes back many millions of years. Way back then, the geological processes took place that provided a sea of oil off the Norwegian coast, which has made this small country (only four million inhabitants!) very well off.

Long after that oil was created, Norway produced one of the most brilliant mathematicians of the nineteenth century: Niels Henrik Abel

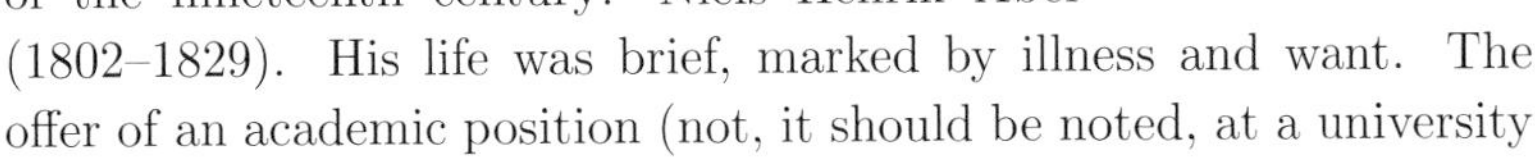

(1802–1829). His life was brief, marked by illness and want. The offer of an academic position (not, it should be noted, at a university

in his native Norway, but in Berlin) came too late. He was too ill to accept it.

It was only after his death that he was recognized in his homeland as a mathematical genius. To honor his memory, the Abel Prize was created in 2002. It is awarded every year to a mathematician whose life work has had a particularly great influence on the development of mathematics. The prize has a monetary value of close to one million dollars, which is about the same level as the Nobel Prizes.

The first prize, awarded in 2003, was given to Jean-Pierre Serre. Subsequent recipients have been Sir Michael Atiyah and Isidore Singer (2004), Peter Lax (2005), Lennart Carleson (2006), and Srinivasa S. R. Varadhan (2007). And Berlin is also involved: the Norwegian embassy has generously financed a trip to the award ceremony for the winning team of Berlin's "Day of Mathematics," which is an annual event for students.

Abel and the Equation of Fifth Degree

Abel made significant contributions in several areas of mathematics. As an example, we shall describe here his work in connection with the solution of polynomial equations.

The problem. Many problems in mathematical applications reduce to the task of finding all numbers that satisfy an equation of the form $x^2 - 2.5x + 3 = 0$ or $x^7 - 1{,}200x^6 + 3.1x - \pi = 0$. (For example, this is everyday work for engineers. From the location of the solutions one can tell whether a system will remain stable or be sensitive to slight disturbances.) The functions that appear in this context (that is, $x^2 - 2.5x + 3$ and $x^7 - 1{,}200x^6 + 3.1x - \pi$) are called *polynomials.* The general polynomial can be written as

$$a_n x^n + a_{n-1} x^{n-1} + \cdots + a_1 x + a_0,$$

where n is some natural number, and the *coefficients* $a_n, a_{n-1}, \ldots, a_0$ are arbitrary numbers.

The largest exponent that appears is called the *degree* of the polynomial. In the two examples above, the degrees are 2 and 7, and the general polynomial $a_n x^n + a_{n-1} x^{n-1} + \cdots + a_1 x + a_0$ has degree

n. We assume that a_n is not equal to zero. (If a_n were equal to zero, we could simply omit the term $a_n x^n$.)

Positive results. It was not until the nineteenth century that it was established that every polynomial $a_n x^n + a_{n-1} x^{n-1} + \cdots + a_1 x + a_0$ has solutions. Since an equation such as $x^2 + 1 = 0$ has no solutions among the real numbers, it is necessary to look for solutions among the complex numbers (and then $x = \pm i$ are the solutions to $x^2 + 1 = 0$). Once complex solutions are allowed, it turns out that the general polynomial $a_n x^n + a_{n-1} x^{n-1} + \cdots + a_1 x + a_0$ has solutions even when the coefficients are themselves complex numbers.[1] But just because solutions exist doesn't mean that one can find a simple formula to express them. In fact, such formulas exist in general only when the degree of the polynomial is "very small." Here are the few positive cases:

- **degree = 1:** Here the problem is to find a number x such that $a_1 \cdot x + a_0 = 0$, where a_1 and a_0 are prescribed constants. The solution is something that every first-year algebra student should know: the equation is easily solved for x, namely, $x = -a_0/a_1$.
- **degree = 2:** Now we need to find all values of x for which the equation
$$a_2 \cdot x^2 + a_1 \cdot x + a_0 = 0,$$
with prescribed values of a_2, a_1, a_0, is satisfied. The solution has been memorized by generations of schoolchildren under the rubric "quadratic formula": the solutions are
$$x_1 = \frac{-a_1 + \sqrt{a_1^2 - 4a_2 \cdot a_0}}{2 \cdot a_2}$$
and
$$x_2 = \frac{-a_1 - \sqrt{a_1^2 - 4a_2 \cdot a_0}}{2 \cdot a_2}.$$
- **degree = 3:** In this case as well, the solutions can be given by an explicit formula. It is called *Cardano's formula*, after the celebrated Italian mathematician Girolamo Cardano, who published the formula in his 1545 work *Ars Magna*.

[1] Cf. Chapter 94.

Beginning with a third-degree equation, one uses a transformation of variables to bring it into the form

$$x^3 - ax - b = 0.$$

Then a solution is given by

$$x = \sqrt[3]{\frac{b}{2} + \sqrt{\left(\frac{b}{2}\right)^2 - \left(\frac{a}{3}\right)^3}} + \sqrt[3]{\frac{b}{2} - \sqrt{\left(\frac{b}{2}\right)^2 - \left(\frac{a}{3}\right)^3}}.$$

- **degree = 4:** In this case as well there are closed formulas for the solutions, and again, one needs only the symbols $+$, $-$, $\cdot$, $\div$ and the extraction of roots to form certain—rather complicated—expressions in terms of the coefficients. The formula was discovered by Ludovico Ferrari (1522–1565), a contemporary of Cardano.

And so on? Why shouldn't it be possible to discover ever more complicated closed formulas for the solutions of equations of ever higher degree? An intense search that commenced in the sixteenth century ended only in the nineteenth, when Niels Henrik Abel settled the matter once and for all.

Abel's Impossibility Theorem. Abel showed in 1824 (at the tender age of 22) that more results of the type given above for equations of degree 2, 3, or 4 are not to be expected. Indeed, for equations of the fifth degree it is impossible to find such a formula—no matter how complicated—that expresses the solutions in terms of the coefficients.

Since that time, mathematicians have known that in many cases, the most they can hope for is to find arbitrarily good numerical approximations to the desired solutions.

Chapter 73

Chance as Reckoner: Monte Carlo Methods

The town of Monte Carlo is well known as a place for automobile racing, celebrity spotting, and casino gambling. Mathematicians, thinking of the vagaries of Lady Luck at the roulette table, have coined the term "Monte Carlo method" to describe techniques in which chance plays a role in computations.

As an example, consider a complex surface F lying inside a square with side length 1. What is the area of F? The classical approach would be to divide the surface into small regions whose areas can be readily computed and then sum up the individual areas.

A Monte Carlo method would take a different approach altogether. Its most important component is a random-number generator that produces a random point in the square. It is important that the generator be programmed in such a way that each point in the square has an equal probability of being generated. The operative expression is "equidistribution." Today's computers can generate millions of such points per second. With this setup, the probability of such a randomly chosen point lying in the region F is proportional to its area.

The Monte Carlo method makes an experimental determination of the area of the region. For example, if out of a million points generated, exactly 622,431 of them lie inside F, that means that the probability of a "hit" is about 62.2%. Therefore, the area of F should be about 62.2% of the entire area of the square, which is 1, and so the area of F has been experimentally computed to be 0.622.

This method has advantages and drawbacks. The main advantage is that Monte Carlo methods can be readily adapted to extremely complex situations: the associated computer program is easily written, since the most important building block, the random-number generator, is built into most of today's programming languages. Unfortunately, chance is unreliable. It could happen that the generated points are not equally distributed over the square, in which case the computed probability of a "hit" fails to correspond to the true area of the region F.

As a result of this defect, the results of a Monte Carlo procedure should be interpreted with caution, along the lines of, "with 99% probability, the area of F lies between 0.62 and 0.63."

It is therefore no wonder that mathematicians look for exact procedures whenever such are available. Or would you rather drive over a bridge whose stability can be guaranteed with only 99% certainty?

A Monte Carlo Method for Computing the Area of a Parabola

As an example of a typical Monte Carlo computation let us calculate the area under a parabola: what is the area between a parabola and the x-axis in the range $x = 0$ to $x = 1$?

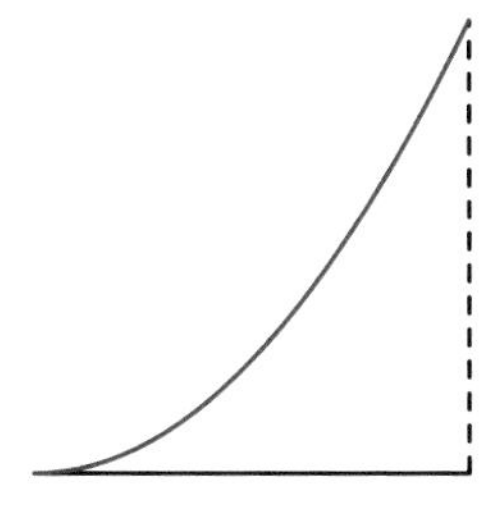

Figure 1. What is the area under the parabola?

This problem is easy to solve exactly. Over two thousand years ago, Archimedes figured out the formula, and today it is the stuff of elementary calculus: Suppose the parabola is given by the equation $f(x) = x^2$. An antiderivative is $x^3/3$, and inserting the upper and lower limits yields an area of $\frac{1}{3}$.

With a Monte Carlo method we can forget about calculus. There are two ways to proceed:

Method 1: Draw the region F whose area is to be determined inside a rectangle R. In this case we can choose a square with side length 1. Then we let the computer generate "many" random points inside this rectangle, in such a way that no point of the rectangle is "hit" with greater probability than any other point. We then have only to count how many of the points landed inside the region F. Since the points are equally distributed, the proportion of "hits" will express the relationship between the area of F and that of R. An example is shown in Figure 2.

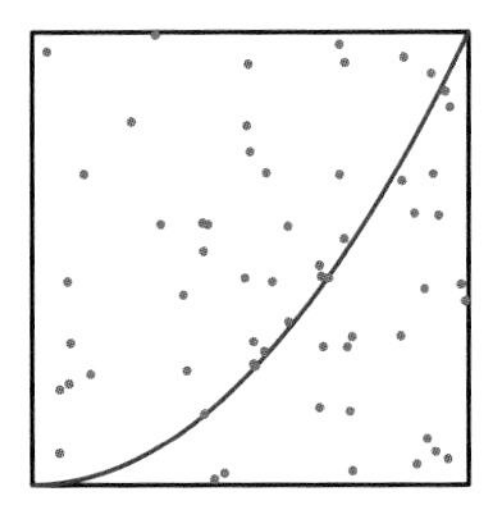

Figure 2. Calculation of area using a Monte Carlo method.

In this little example, sixty points were generated, of which 22 lay within the region under the parabola. Therefore, the area of the region should be approximately equal to 22/60, which is $0.366\ldots$.

The result is not at all bad for so few points. For a computer it is no problem to generate many times that number of points and thereby increase the accuracy and reliability of the result.

Method 2: This method rests on a probability-theoretic interpretation. The area sought is equal to the average payout in a game in which points in the unit interval $[0, 1]$ are chosen at random and a payment of x^2 is received. To make use of this fact, we can write the following computer program: Set a memory register r to zero and let the computer produce numbers at random between 0 and 1. Each number is squared and added into register r. (For example, if the random number 0.22334455 is generated, then the value of r is increased by $0.22334455 \cdot 0.22334455 = 04988278801$.) One does this many, many times and then divides the result by the number of random numbers generated (which we shall call n here). Using standard programming "pseudocode," we get the following program:

```
⋮
n := 10,000;
r := 0;
for i = 1 to n do
  begin y := random; r := r + y * y; end;
r := r/n;
⋮
```

When the program has finished execution, the register r contains an approximate value for the area under the parabolic segment. Here is the result of several computer simulations:

Number n of Trials	10,000	10,000	100,000	100,000
Register r	0.3338399	0.336283	0.33350	0.33304

It is remarkable that one can get a rather good approximation to the true value 0.3333... without using integration. It takes only a fraction of a second, and the function being computed could of course be much, much more complicated. The only drawback is that one can never be one hundred percent certain. Only if one knows already in advance what the outcome should be can one assess whether the approximation is a good one. If one does not have this information, one must trust the computer and the laws of probability, and in the case of calculations on which lives and property depend, one would want to think twice before relying on the result.

Chapter 74

Fuzzy Logic

In the recent past, vacuum cleaners and washing machines were advertised as operating with "fuzzy logic." The idea of this form of logic, proposed in the 1970s by the Berkeley mathematician and computer scientist Lotfi Asker Zadeh, was to give a mathematical basis to the way we reason in everyday life.

If one insists on precision in mathematics, then there is only "true" or "false." An integer is either a prime number or not prime, and there is no gray zone between the two concepts.

But our daily lives are not at all black and white. Depending on the available information, we often have only a rather vague idea about whether some statement might be true: Is this mode of transportation safe? Is this business venture worthwhile?

Fuzzy logic attempts to "humanize" mathematics by allowing not only "true" and "false" as truth values of a proposition, but all values between 1 (absolutely certainly true) and 0 (absolutely certainly false). A value of 0.9, say, could be applied to a statement about whose truth one was "quite certain."

Remarkably, large portions of classical logic can be brought over into the fuzzy realm. For example, fuzzy statements can be joined: if the propositions p and q have high truth values, so does "p and

q." This models our experience quite well, and therefore many mathematicians have found it attractive.

Similar approaches can be used to control more or less complex processes. Suppose a bar is to be balanced on a horizontal plate controlled by a robot arm. The precise modeling of the situation is extremely delicate, but a fuzzy-logic implementation can easily be accomplished. One can associate fuzzy-logic values for the degree of the bar's divergence from the perpendicular with statements such as "the bar is rotated a bit to the left," "the bar is rotated a lot to the left," and so on. If the displacement is, say, ten degrees to the left, the fuzzy value for "a bit to the left" could be 0.6, while the value for "a lot to the left" might be 0.4. (Since we are more confident that at ten degrees, the bar is rotated a little to the left than a lot, its fuzzy value is larger.) Furthermore, one must prescribe what reaction is expected from the robot in relation to "a bit to the left" and "a lot to the left." It could be "move the plate one inch to the right" and "three inches to the left," for example. After the motion is observed, an evaluation is made: the "truer" a result is estimated to be, the greater is the portion of the associated reaction that is executed.

With this procedure it is also possible to make use of human knowledge, in particular that which cannot be expressed in mathematical terms. However, for most mathematicians, fuzzy techniques are a makeshift device. They would much rather have a precise logic in their vacuum cleaners, even if one doesn't perhaps see this in the end result.

Fuzzy Control

Classical control theory is a difficult mathematical discipline. Important contributions to the field were made by the American mathematician Norbert Wiener (1894–1964), who coined the word *cybernetics*. The idea is to control a system optimally: certain goal values should be reached as quickly (or as cheaply) as possible, and during the process, one can influence the course of events through control parameters. The "system" might be a chain of chemical reactions in a pharmaceutical factory, a blast furnace, or an enemy rocket that is to be shot down. The range of applications of the methods developed in this area is enormous. Things can become very complicated because one may not be in possession of complete information, because only a greatly delayed influence is possible, or because random developments during the process can alter the course of events in an unpredictable way. Normally, very complex equations represent the control functions, and an exact solution is possible only in exceptional cases.

With *fuzzy control* one can make one's life much easier, as we have indicated in this chapter. The system is observed, and then an evaluation is made as to the extent to which the current situation matches various scenarios. If the bar to be balanced is leaning five degrees forward, then that could lead to the following distribution of the scenarios mapped out previously (from "strongly leaning backward" via "leaning somewhat backward," "not leaning," "leaning somewhat forward," to "strongly leaning forward"): $0, 0, 0.2, 0.8, 0$. Then "experts" are asked, what should be done in the case of strongly leaning backward? leaning somewhat backward? And so on. If one discovers, among other things, that one should do nothing if there is no leaning,[1] and with a somewhat leaning forward one should move the plate five inches forward, one combines these two reactions in proportion to their participation in the given scenarios: with proportion 0.2 one does nothing, and with 0.8 one pushes the plate five inches forward. The result is a movement of $0.8 \cdot 5 = 4$ inches.

Remarkably, one can handle quite complex control problems in this way. The result might be somewhat more disjointed than a classical solution, but in return, "fuzzification" is much easier to implement.

[1] Of course, you probably figured that out yourself without consulting an expert.

Chapter 75

Secret Messages in the Bible?

For mathematicians, numbers are objects whose properties are to be investigated and that are to be used as aids in computation. They do not ascribe to numbers any sort of mystical properties. Yet there is a long tradition, going back at least to Pythagoras, of seeing something deeper in numbers. For example, numbers can be associated with certain qualities ("twoness is the font of change," "threeness speaks of wisdom and knowledge"), and then one can use such relationships in decision-making.

Should one buy a house if the sum of the digits of the street address is an "unlucky" number? Similarly, one can allow oneself to be influenced by the registration number on a used car (buy or not buy) or the birth date of a prospective spouse: numbers are everywhere.

In the nineteenth century, number mysticism was particularly widespread. A popular pastime was to assign numbers to letters of the alphabet and then assign a numerical value to individuals' names. If the result was 666, a number against which one had long been warned in the Bible as the "number of the beast," that must certainly mean something:

> Here is wisdom. Let him that hath understanding count the number of the beast: for it is the number of a man; and his number is Six hundred threescore and six (Revelation 13:18).

The weak point of the method is that there are so many ways of associating numbers to letters that one has great latitude in influencing the outcome. If the result isn't quite what one wanted, one can simply manipulate the way the name is written. In Tolstoy's *War and Peace*, for example, Napoleon can be associated with the number 666 only when his name is pronounced slightly incorrectly as "Le Empereur."

Another variant of number mysticism made a big splash in the year 1997 when the book *The Bible Code*, by M. Drosnin, appeared, which offered the theory that in the original Hebrew Bible there was a great deal of encrypted information about past and future events.

The debate reached even the professional mathematics journals, for it wasn't at first clear how there could have been so many accurate predictions just waiting to be decoded. But it turned out that Drosnin's method could be made to yield similar results in any text of sufficient length. One had only to search long enough.

Of course, such games are not confined to the Bible; there are contemporary variants as well. Microsoft critics can, if they wish, transform "Bill Gates" into a 666: one must simply write the number "correctly" as "B. & Gates" and transform the letters into computer ASCII code:

	B	.	&	G	A	T	E	S	sum
ASCII value	66	190	38	71	65	84	69	83	**666**

But there is more than one way to skin Bill Gates:[1] Since the man's full name is William Henry Gates III, one can see what happens with "BILL GATES 3," and sure enough:

	B	I	L	L	G	A	T	E	S	3	sum
ASCII value	66	73	76	76	71	65	84	69	83	3	**666**

[1]The following appeared in *Harper's* in 1995.

Admittedly, there is a bit of cheating here. First of all, the number 3 does not have ASCII code 3, but rather 51. And secondly, the space between the first and last names is missing (ASCII code 32). A direct translation of his name would never have exposed Bill Gates as a scoundrel.

Number Mysticism Begins with Pythagoras

The history of number mysticism can be traced to ancient times. It is first found among the Pythagoreans, the disciples of Pythagoras (about 500 B.C.E.). In the earlier Egyptian and Babylonian civilizations, numbers were important tools for carrying out important calculations, such as in astronomy and architecture, and no further meaning was ascribed to them. Nor did the Greek mathematicians two hundred years after Pythagoras have any room for number mysticism; indeed, in the great compendia of mathematics, such as Euclid's *Elements*, there is not a word on the subject.

After the decline of the Pythagoreans, number mysticism was almost forgotten, until it was taken up again by the Neo-Pythagoreans around the beginning of the common era. And since then it has held a secure place in the reservoir of the irrational. Particularly in evil times, when people are searching for explanations for their troubles and help with their lives, and their religion seems insufficient, it can suddenly once again become important that 1 is a "good" number and 2 a "bad" one.

Although all of this has played not the slightest role for mathematics, one should not believe that scientists are always immune to worldviews that today are largely seen as irrational. Johannes Kepler, to whom we owe the knowledge that the planets move in elliptical orbits, attempted to explain the distances of the planets from the Sun in reference to a set of nested Platonic solids (cube , tetrahedron, octahedron, icosahedron, dodecahedron). And the great Newton apparently spent more time in his alchemical laboratory and searching for secret messages in the Bible than in the writing of his *Principia Mathematica*, the book that inaugurated the triumphal march of mathematical methods in the natural sciences.

The "Law of Small Numbers"

"Mystical" connections sometimes arise simply because the number of small numbers is small. The mathematician Richard Guy dubbed this situation the "law of small numbers."

The mathematical reasons are indisputable. Anyone wishing to place five balls in four boxes has no choice but to place more than one ball in at least one box. Mathematicians call this the *pigeonhole principle* (more on the role of this principle as a method of proof can be found in Chapter 61). By this principle, it is unavoidable that in associations among related concepts, the same number will occur quite often. For example, we might have the following groups of three:

- three graces (Aglaia, Euphrosyne, and Thalia),
- three kings (Caspar, Melchior, and Balthazar),
- three musketeers (Athos, Porthos, Aramis),
- three tenses (past, present, future).

The significance of this coincidence is nil, but numerologists find the occurrence of groups of three to be full of meaning.

Anyone interested in reading more on the subject may wish to consult Underwood Dudley's *Numerology: Or, What Pythagoras Wrought*. Spectrum, 1997.

Chapter 76

How Knotted Can a Knot Be?

Imagine that in your closet you have a long extension cord. You plug it into itself—the plug at one end into the receptacle at the other—and thereby create a closed circuit.

If the cord happened to be more or less tangled up before you plugged it into itself, it is now hopelessly knotted (see Figure 1). Is it possible to undo the knot without unplugging the cord? That is, can it be unknotted into a large circle? It is clear that sometimes this is possible, and sometimes not.

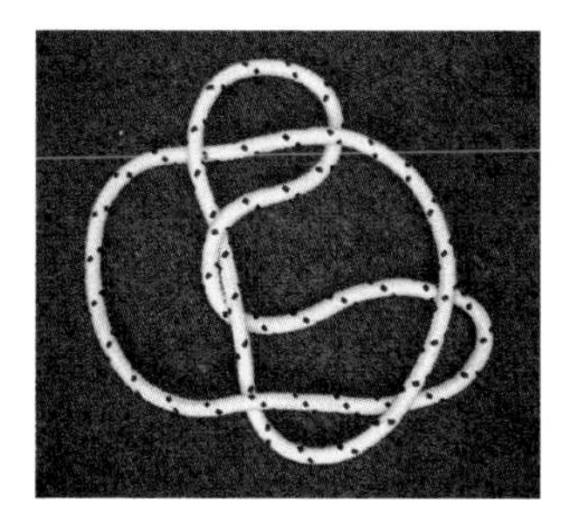

Figure 1. Can this tangle be unknotted?

But when exactly? This question has engaged mathematicians for several centuries. Of course, it is not a theory of extension cords, but of general abstract knots. The first problem that must be dealt with is to find a suitable vocabulary to describe questions about knots. This issue was raised by Leibniz, but it was solved satisfactorily only at the end of the nineteenth century. The precise formulation is somewhat technical, so we shall stick to extension cords.

Scandalously, it took another several decades before one of the simplest questions about knots was resolved. It is a fact known by everyone who has ever tied a knot, but it was proved rigorously only in the 1930s: there are some knots that cannot be undone, even with the most refined methods. Here is the simplest example, the cloverleaf knot:

A much more difficult task is the *classification problem*: how can one categorize the various truly different types of knots? This issue is the subject of current research.

The principal motivation for the development of knot theory comes from its significance for physics. In 1867, the English physicist William Thomson, later Lord Kelvin, proposed a new and original atomic theory, according to which an atom was to be modeled as a vortex line in the ether; one could imagine atoms as interlocking smoke rings. The variety of possible atoms would then correspond to the different basic types of possible knots. Thus arose the classification problem and a systematic investigation of the theory of knots.

Kelvin's ideas have not found a place in modern physics. However, in the meantime, knot theory has become of vital importance to physics for another reason altogether. It plays an important role in *string theory*, which attempts to describe the basic structure of matter.

Knot Invariants

It took 230 years from the formulation of the question, "are there knots that cannot be unknotted?" to its first solution by the mathematician Kurt Reidemeister (1893–1971). In 1932 he proposed a solution based on the use of *knot invariants*.

Let us first explain the idea of working with invariants with a simple example:

> Consider the following simple "game." Ten stones are placed on a table, and a turn consists in either placing seven stones on the table (taken from an unlimited supply) or removing seven stones (if possible).
>
> *Problem:* Is it possible for there to be exactly twenty-two stones on the table at some point in the game?
>
> *Solution:* No. It is impossible, and one can prove this with the following simple invariant technique. At every point in the game we consider the remainder when the number of stones is divided by seven.[1] Then three things are clear:
> - The remainder at the beginning of the game is 3.
> - A turn in the game does not alter the remainder, since the number of stones always grows or shrinks by 7.
> - For the number 22, the remainder on division by 7 is 1.
>
> It follows that the number 22 can never be achieved.

And now back to knot theory. Reidemeister had the idea to use an analogous approach for knots. He first defined what we might call a "simple game move" for knots. He described three different types of "Reidemeister moves." These are manipulations along the lines of "move one loop completely over another loop." What is crucial here is the observation that any manipulation that one can perform on a knot can be represented as a sequence of Reidemeister moves.

[1] That is, in the language of Chapter 22, the number of stones modulo 7.

And then Reidemeister defined an invariant: there is a property of knots that is not altered by performing these moves; if a knot has this property before a move, it will have it after the move.

This invariant is, alas, much more complicated than "the remainder on division by 7" from our illustrative example above. The Reidemeister invariant is the possibility of coloring a diagram of the knot in the plane in a specified way.

The point of Reidemeister's approach is that one can prove the following assertions:

- The invariant does not change on the performance of a Reidemeister move.
- A closed circle, that is, the unknotted knot, cannot be colored in the way specified.
- Certain knots, such as the cloverleaf knot shown earlier, can be so colored.

It therefore follows that the cloverleaf knot cannot be untied.

We should note that this result by no means answers all of the outstanding questions. *If* a knot is colorable, then it cannot be untied. But the converse of this statement is false: in many cases this technique cannot determine whether a knot can be unknotted, since there are knots that while not colorable, cannot be unknotted. The challenge is then to find new invariants that will distinguish such knots from those that can be untied.

The search for such invariants is a topic of active research. A long-term goal is a *universal invariant*, that is, a property of knots that is easy to check and that a knot satisfies if and only if it can be unknotted. But finding such an invariant seems to be a long way away.

Chapter 77

How Much Mathematics Does a Person Need?

How much mathematics is really necessary? Do we really need quadratic equations, graphs of functions, and integrals? Aren't addition and the multiplication table enough, so that we can estimate how much we have to pay at the grocery store and figure the tip at a restaurant? Some would go even further and advocate the relegation of all such computations to the pocket calculator, which will certainly soon be built into every cell phone.

One shouldn't take such radical proposals too seriously, for one could just as well advocate the demise of other school subjects, such as English (spell- and grammar-checking programs) and geography (Google). Nonetheless, one may justifiably ask what the proper place of mathematics is in today's educational system.

In my view, there are *three points* that justify a more than superficial involvement with the subject. First, it is undisputed that mathematics is *useful* for solving concrete problems in the real world. It begins with mental arithmetic at the baker's and continues into essentially every branch of science. Any student thinking of a career in the natural sciences, engineering, the humanities, social sciences, or medicine will need solid grounding in at least the fundamentals of statistics. Computers are no substitute, no matter how user-friendly they may be. Anyone who is incapable of doing a simple addition to verify that the computer output makes sense will also not notice when the cashier at the supermarket accidentally punches a price into the cash register with the decimal point in the wrong place. And even the most advanced statistics software package does not absolve the user of the responsibility of checking whether the procedure being used is valid for the intended application, of determining what questions can validly be asked of the accumulated data, and deciding how the results are to be interpreted. Without mathematics, one is at the mercy of the panic mongers and profiteers, and long-term economic planning such as the purchase of a home becomes a risky game of chance if one is unable to assess the amount of actual debt being assumed.

Second, mathematics can exert the utmost *fascination* as an intellectual discipline. Problem-solving requires perseverance and creativity, qualities the cultivation of which can never be given too much attention. Personnel directors of major corporations are fond of saying that these qualities, to be found in students of mathematics, are at least as important as technical knowledge of the firm's business. A mathematician is used to chewing over a problem until a solution has been obtained. There is no disputing that such a quality can be valuable in any profession.

And third, it should not be forgotten that *our world is built on mathematical principles*. Since Galileo we have known that the book of nature is written in the language of mathematics. Thus all those wishing to know the innermost nature of the world must have knowledge of numbers, geometric objects, and probability.

Mathematics therefore plays an important role in all basic questions regarding the natural sciences. A philosopher can hardly dare to attempt to speak intelligently about ontology without the mathematical prerequisites for understanding relativity and the theory of probability.

Unfortunately, in the schools, instruction frequently gets no further than the technicalities. A student can achieve good grades by memorizing cookbook recipes, and anyone whose knowledge of mathematics remains at this level has missed what is essential. It is as though one studied French, say, by learning only the grammar and never read a poem by Baudelaire. But that is another story.[1]

Read All about It

Most of the chapters in this book can be read as illustrations of the three aspects of the importance of mathematics discussed in this article. Here are some examples:

- "Mathematics is useful": Chapters 1, 7, 9, 14, 21, 62, 63, 64, 71, 90, 91, 93, 98.
- "Mathematics is fascinating": Chapters 4, 15, 17, 18, 23, 33, 48, 49, 76, 99.
- "Mathematics is the language of nature": Chapters 38, 47, 51.

[1] See Chapter 31.

Chapter 78

Big, Bigger, Biggest

You are at the farmers' market, contemplating two bushel baskets of apples, and you would like to buy the basket that has the greater number of apples in it. How can you decide which basket has more apples? The first solution that comes to mind is to count the apples in both baskets and compare the results.

But what if you can't count that high? It is still possible to make a determination. Simply remove an apple from each basket repeatedly until one of them is empty. That is the basket with the smaller number of apples.

In a similar way, one can compare the sizes of sets, even if one is unable to count their members. This idea was applied successfully by Georg Cantor, the founder of set theory, even to sets that are infinitely large. One has only to effect a slight modification to the apple example. Instead of removing pairs of apples, one arranges the apples in the first basket in a long row. Then one attempts to place those in the second basket in a parallel row, one apple from basket two beneath each apple from basket one. A perfect matching indicates that the two baskets have the same number of apples, and if the numbers are not the same, that will be noticed immediately.

In this way we may deduce that there are "exactly as many" odd numbers as there are even numbers. One has simply to bring the sets of numbers $2, 4, 6, \ldots$ and $1, 3, 5, \ldots$ into alignment, which can be done by associating the even number 2 with the odd number 1, the even number 4 with the odd number 3, then 6 with 5, and so on, each even number being associated with the odd number that is smaller by 1.

Cantor made a number of surprising discoveries regarding phenomena relating to the sizes of sets of numbers. As an example, let us consider the set of rational numbers, that is, numbers that can be represented as fractions, or quotients of two integers, such as 7/9 and 1,001/4,711. This set has the same number of elements as the set of natural numbers $1, 2, 3, \ldots$. It is not at all obvious that such should be the case. It would seem at first glance that there should be many "more" fractions than whole numbers.

Cantor was also able to show that the set of fractions is tiny in comparison to the set of all numbers. The set of all numbers representable by infinite decimals is so vast that no method can exist by which all the elements of this set can be associated with the numbers $1, 2, 3, \ldots$.

In many ways, infinities behave like ordinary numbers. For example, they can be compared, so that for two infinite sets, either they have the same size, or else one is bigger, "more infinite," than the other. The arithmetic of infinite sets is full of pitfalls, with paradoxes and false conclusions at every turn. And therefore, most of mathematics operates in the realm of sets that are "not too big."

There Are Just as Many Fractions as Natural Numbers

The surprising fact mentioned above that there are just as many fractions as there are natural numbers can be clarified with a picture. First, however, it must be clear that the statement, "the set M has the same number of elements as the natural numbers," means that a walk can be taken through the set M in such a way that every element of M is encountered exactly once. That is indeed the case when, for example, M is the set of even numbers. At the nth step, one visits the even number $2n$. Thus each even number is encountered exactly once, for example, 4,322 at the 2,161st step.

But how is one to walk among the fractions so that one reaches all of them? Cantor had the following idea. The fractions are arranged in the following clever way. On the first line, one writes down all the fractions with denominator 1, like this:

$$0,\ 1,\ -1,\ 2,\ -2,\ 3,\ -3,\ \ldots.$$

The signs alternate, so that all negative numbers are included.

Then in line 2, all the fractions with denominator 2 (when expressed in lowest terms) appear:

$$\frac{1}{2},\ -\frac{1}{2},\ \frac{3}{2},\ -\frac{3}{2},\ \frac{5}{2},\ -\frac{5}{2},\ \ldots.$$

The following lines have fractions with denominator 3, 4, 5, and so on.

Every fraction appears exactly once in this infinite array. For example, 12/1,331 appears in the 1,331st row. Now all that is missing is a walk through this array. A naive approach will not work. If you start walking along the first row, you will never get to the end of row 1 and hence never visit $\frac{1}{2}$ and everything beyond. The trick is to move along the diagonals, as shown in Figure 1.[1]

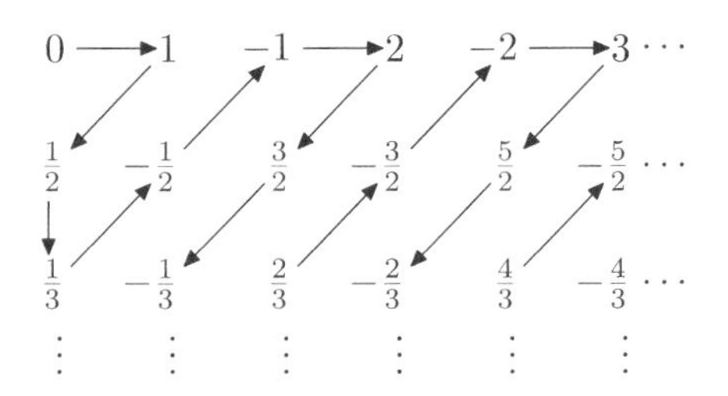

Figure 1. A "walk" that meets every fraction.

Starting at zero, the subsequent steps are $1, \frac{1}{2}, \frac{1}{3}, -\frac{1}{2}, -1, \ldots.$ Even though it is not easy to specify where the walker would land, say, on the ten thousandth step, it is nonetheless clear that every fraction will be reached at some step. And therefore there are just as many natural numbers as there are fractions.

[1]This proof goes under the name *first Cantor diagonal method.*

Chapter 79

It Is Probably Correct

In recent decades it has become ever more apparent that randomness plays more of a role than causing unpredictable disruptions. We have already written about Monte Carlo techniques in Chapter 73, which allow one to delegate complex calculations to random processes. Today we are going to talk about something even more fundamental: randomness can help in discovering the truth.

For example, consider a large number n, say of several hundred digits. It can be important, for example in cryptographic applications, to know whether the number n is prime. On account of the size of the number, direct methods are infeasible, and one needs to take another point of view.

It is known from the theory of numbers that if n is not prime (i.e., composite), then at least half the numbers between 1 and n have a certain property P related to n that is easily verified, while if n is prime, then no numbers at all have the property. (The exact nature of P is unimportant for our discussion.) One can now test n for primality by using a random-number generator to select numbers x between 1 and n and testing whether they have property P. If a number x tests negative, then that is either because n is in fact prime or because while n is composite, x happened to be among the portion of numbers that don't have property P. It is quite unlikely that a large number of tests would come up negative unless n were in fact prime. After twenty negative tests, the likelihood that n is not prime is about 2^{20}, that is, less than one in a million.

In this way, mathematicians arrive at statements of the type, "this number is prime with overwhelming probability." In many applications, such "industrial-grade primes" are wholly satisfactory. And if one feels insecure about a "prime" that might be composite with probability one in a million, then you can perform forty of the above tests instead of twenty and have a "prime" that is composite with probability only one in a trillion. Such numbers, while not known with mathematical certainty to be prime, are considered prime with a moral certainty that allows them to be used with confidence in many applications.

In fact, in many cases it is not even necessary to know the correct result with overwhelming probability. If a technique allows you to crack a secret code with fifty percent probability, then those using that code will be shaking in their boots. Similarly, would you sleep peacefully in your bed at night if there were a bunch of keys hanging from your front door half of which could open the lock?

It should be emphasized that in most applications it is better to stick with the classical methods in which a solution can be calculated exactly. Indeed, would you willingly go up to the the ninety-fifth floor of a building that has been "proven" by the structural engineers to be stable with a probability of 99 percent?

Breaking Secret Codes: With High Probability

In this connection we should mention the algorithm of Peter Shor that would be able to factor a number that was the product of large primes if there were a quantum computer on which to run it (cf. Chapter 23). We observe once more that the difficulty believed to be inherent in factoring large numbers is crucial for the security of current cryptographic methods,[1] and therefore Shor's algorithm dropped like a bombshell on the world of cryptography.

Suppose, then, that p and q are large prime numbers, and let n denote their product: $n = p \cdot q$. Now a random number x between 1

[1] This is discussed more fully in Chapter 23.

and n is generated, which any classical computer can do with lightning speed using a built-in random-number generator. It is known that the factors of n could be determined if one knew a certain characteristic quantity associated with x called the *period*, which has a certain property P in at least half the cases. Then a quantum computer could be programmed to compute the period with very high probability. Here is the algorithm:

(1) Generate a random number x between 1 and n (that is accomplished quickly on current computers).

(2) Compute a candidate for the period of x on the quantum computer (that would also be quick if quantum computers actually existed).

(3) Repeat step (2) until the period of x is actually found. (This can be rapidly tested on today's computers.)

(4) Check whether the period has the property P (which can also be delegated to computers existing today). If the test is negative (the number does not have property P), return to step (1) and choose another random number x.

(5) Use x to determine p and q, and then deciphering the code is no problem.

Note that here chance plays an important role *at two places*. First, the period is determined only with a certain probability, and second, only half (at least) the numbers x are suitable for determining the factors of n. For this application that is not a serious drawback, since it is unimportant whether it takes a few minutes more or less to decrypt an encoded message.

Not much has been heard about quantum computers in recent years. First of all, no one has any idea how to overcome the enormous technical problems that stand in the way of truly interesting applications to cryptography. And second, it has turned out to be surprisingly difficult to translate interesting problems in such a way that a quantum computer would be able to obtain a solution with high probability.

Chapter 80

Is the World a Crooked Place?

One of the high points of mathematics occurred more than two thousand years ago, when Euclid produced a systematic compilation of the fundamentals of plane geometry. What are the relationships between the angles produced by a line intersecting a pair of parallel lines? What is the sum of the angles of a triangle? Of a trapezoid? What does it mean to construct with straightedge and compass?

The precision of the approach is impressive, but the content is anything but surprising. It is clear to anyone that there is only a single straight line through any pair of points and that for any line L and a point P external to L, there is a unique line through P that is parallel to L.

In other words, Euclid's axioms amount to making precise our everyday experience and codifying it mathematically, and therefore his geometry remained unquestioned gospel until a mere 150 or so years ago.

But then in the nineteenth century, mathematicians began to question the Euclidean worldview. The great Carl Friedrich Gauss (see Chapter 25), for example, performed an experiment on an enormous triangle spanning three mountain peaks (Brocken, Inselsberg, and Hoher Hagen) to determine whether on Earth, the angles of a triangle indeed add up to 180 degrees. Within the accuracy of his measurements, Euclidean geometry gave the correct answer, but it is noteworthy that Gauss considered it necessary to compare theory with reality.

In the 1830s, Bolyai and Lobachevsky developed, independently of Gauss and each other, a theory of non-Euclidean geometries. These are constructed formally, along the lines of Euclidean geometry. However, in these geometries, triangles do not necessarily have an angle sum of 180 degrees. Then in the 1850s, Bernhard Riemann developed the theory further, giving a very general model for abstract geometries.

For several decades, these ideas were known only to a few specialists, but they reached a much wider audience when Einstein's general theory of relativity demonstrated that the structure of the universe can best be modeled by a geometry à la Riemann. If the universe were two-dimensional, then one could imagine it as an undulating surface, whose curvature at any particular spot is a measure of the amount of mass present at that location.

This rather abstract theory has in the meantime been verified experimentally. The curvatures that appear are extremely small, smaller than the error that Gauss's measurements would have produced in his mountain triangle experiment. But even so, these tiny differences between Euclidean and Einsteinian geometries are often of significance to you and me in our daily lives. For example, the synchronization of the satellites used in the Global Positioning System (GPS) actually depends on the general theory of relativity.

A Triangle with an Angle Sum of 270 Degrees

I would like to emphasize that in making his measurements, Gauss was indeed working with a triangle whose sides were straight lines connecting the various mountain peaks. After all, he was taking his measurements by eye, and since light travels in a straight line in a homogeneous medium, the result is a triangle with straight sides.

However, one can measure large triangles on Earth in another way. One can imagine a triangle as laid out along the Earth's surface, determined by three points and the shortest distance between each pair of points. Of course, we allow only paths that travel along the surface of the Earth; no shortcuts through the Earth's crust are permitted.

The shortest lines between two points on a sphere are called *great circles*. They connect points along a circular arc whose center is the center of the sphere. (Perhaps you have wondered why it is that a flight from New York to Hong Kong goes over the North Pole and not, say, over Hawaii. A great circle joining those two cities comes very close to the North Pole.)

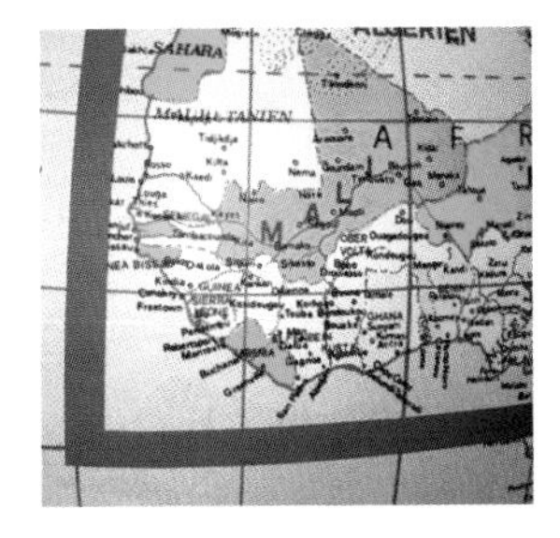

Figure 1. A right triangle and a right triangle on the Earth's surface.

If we interpret great circles as lines, then we end up with a *spherical trigonometry*, which exhibits some phenomena that take a bit of getting used to. For example, one can easily draw a triangle in which all three angles have a measure of 90 degrees (see Figure 1). Thus in contrast to the geometry of the plane, on the sphere, the sum of the angles of a triangle can equal 270 degrees. To draw such a triangle, one could start at the North Pole, travel along a great circle to the equator, then turn east or west (your choice) and travel along the equator a distance of about six thousand miles (to be precise, exactly one-fourth of the Earth's circumference), and then turn north and take a great circle back to the North Pole.

Chapter 81

Is There a Mathematical Bureau of Standards?

In the beginning was the word. Just as in other areas of human endeavor, notation and convention play an important role in mathematics. Why does the number that represents the ratio of the circumference of a circle to its diameter have its own name, π? Why is two raised to the zero power equal to one?

The reasons for such conventions are many. Sometimes, it is the result of a single historical moment; more often, there is a gradual agreement on the most pragmatic approach. For example, it is clear that in working with relationships among the parts of a circle, the number represented by π arises frequently: the circumference is equal to 2π times the radius, where π is the number with decimal expansion $3.14159\ldots$. There is no particular reason why this number, rather than some related number, should have been ennobled with its own special symbol. Much ink would have been saved in the history of mathematics if instead of $3.14\ldots$, twice the value of π, namely $6.28\ldots$, had been given its own name. If that number had been given its own symbol, say ϖ, then we would have the simple relation "circumference is equal to ϖ times the radius." And in fact, in higher mathematics, two times π, that is, our ϖ, appears much more frequently than π by itself. Too late! Trying to institute such a reform in mathematics would have little more chance of success than the universal adoption of Esperanto or George Bernard Shaw's proposals for a phonetic spelling of the English language.

The situation is somewhat simpler in the case of mathematical conventions based on pragmatism. They are simply a result of enlightened laziness. Thus the convention that for any nonzero number a, the zeroth power of a is equal to 1 ($a^0 = 1$) obviates the need for a complicated law of exponents with all sorts of special cases, so that one need memorize and work with a single formula.

In this connection, we might consider the definition of *trapezoid*. To a mathematician, a trapezoid is a quadrilateral with a pair of parallel sides. However, the trapezoids to be found in schoolbooks always show the horizontal sides as parallel (see the left-hand picture of Figure 1) and never the vertical (as in the middle picture). And the line on the top is always shorter than the line on the bottom. And furthermore, such trapezoids always have *exactly one* pair of parallel sides. Yet a rectangle also has a pair of parallel sides; in fact, it has two such pairs. But if rectangles were not to be considered trapezoids, all the results proved for trapezoids would have to be proved all over again for rectangles, and that would be highly uneconomical.

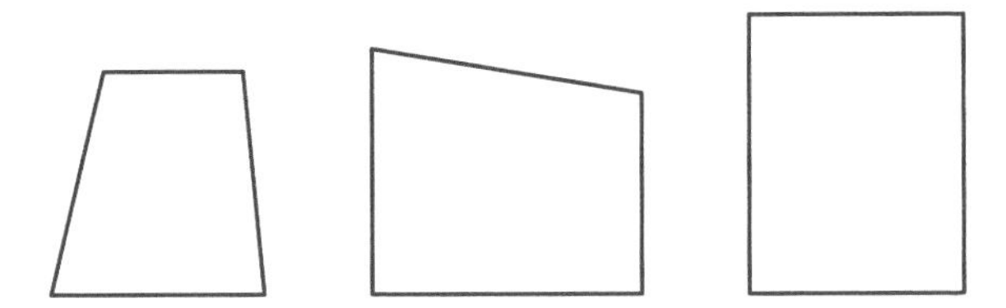

Figure 1. Some trapezoids.

Nowhere in the world is there a mathematical commission on standards. New terminology and notation are proposed, and they are accepted or rejected over time by the mathematical community without much fanfare: most of mathematicians' energy is focused on more-important matters. The give and take involved in the establishment of mathematical convention may be hard to imagine for those who think of mathematics as somehow "given" from on high. In the end, a rectangle is also a trapezoid because of general agreement—that a trapezoid has *at least* one pair of parallel sides instead of *exactly one pair*—and not because God in heaven or some committee of wise persons once upon a time laid down the law.

Why Is 1 Not a Prime Number?

After this chapter appeared as a newspaper column, it was called to my attention that there is, in fact, a group of German mathematicians who would like to introduce some norms for mathematics along the lines of the DIN (*Deutsche Industrie-Normen* = German Industrial Standards). However, even though there may be good grounds for such norms, years have passed and professional mathematicians are largely unaware of such proposals.

The reasons are manifold. First of all, there is the law of inertia: people stick to the terminology and notation that they learned at school. Then there is the problem of common sense versus convenience. For example, if A is a subset of B,[1] is it better to write $A \subset B$ or $A \subseteq B$? The second notation is more sensible, since the notion of subset allows for equality of the two sets, and so the symbol $\subseteq$, being analogous to the "less than or equal" symbol $\leq$ for numbers, is a more rational choice than $\subset$, which corresponds to the "less than" symbol $<$. But reason be damned. The symbol arises so often that $\subseteq$ is generally sacrificed in favor of $\subset$, which saves a lot of lines on the chalkboard. (The German DIN commission has—of course—decreed $\subseteq$.)

Finally, there are problems that might be called ideological. What set the symbol $\mathbb{N}$ represents depends on whether one includes zero among the natural numbers. For many (such as logicians), $\mathbb{N}$ denotes the set $0, 1, 2, 3, \ldots$, but most other mathematicians consider it to be the set $1, 2, 3, 4, \ldots$.

In the end, it doesn't matter which convention is used, but one must read the introductory pages of a mathematics text carefully to know what will be in force in the pages to come. As a rule, one

[1] A set A is a subset of a set B if every element of A is also an element of B. For example, the set consisting of the elements 1 and 3 is a subset of the set consisting of $0, 1, 2, 3$. Note that by this definition, every set is a subset of itself.

chooses the notation that makes for more convenient formulations in one's daily work. If one defines a prime number as a natural number that is divisible only by itself and 1, then the number 1 is included among the primes. But if 1 is included among the primes, the price to be paid is a loss of the assertion that every integer is uniquely representable as a product of prime numbers, since now, for example, one could write 6 as $2 \cdot 3$ as well as $1 \cdot 1 \cdot 2 \cdot 3$, the first time with two "prime" factors, and the second time with four.

Since the uniqueness of prime factorization is an important desideratum in mathematics, 1 must be blackballed out of the club of prime numbers. The definition that has been agreed on by custom and usage is this: a prime number is a natural number greater than 1 that is divisible by only itself and 1. Therefore, throughout the world, from Tuva to Vanuatu, from Dublin to Lublin, two is the smallest prime number, and the uniqueness of factorization is assured.

Chapter 82

The Butterfly That Fluttered By

"A butterfly flapped its wings in Greece, setting off a tornado in Florida." This statement from *chaos theory* has achieved a wide degree of recognition among the general public, with "Greece," "Florida," and "tornado" replaced by other geographical regions and other types of storm. But what does such a statement really mean?

In a rather superficial sense, the statement is of course true, since everything is "somehow or other" dependent on everything else. However, it is impossible to describe this interdependence more precisely, since even an exact description of the wind currents around the butterfly exceeds our capabilities.

What is significant about the butterfly's flight is that it can be used as an illustration of a phenomenon that occurs in many areas of our lives: tiny alterations in the initial conditions of a process can lead to enormous effects on the result.[1] Anyone who has ever played three-cushion billiards knows that slight variations in the angle at

[1] Translator's note: This phenomenon is nicely illustrated in Edgar Allen Poe's tale "The Gold-Bug."

which the cue ball is struck lead to dramatic differences in the final position of the ball.

The consequences of statements like the one about the butterfly are more philosophical than practical. Since we can always know the initial position of a system only to within some unavoidable error, we shall never have remarkable success in trying to see into the future with much detail. The optimistic expectation of Pierre de Laplace (1749–1827), who pictured the world at the beginning of the nineteenth century as a vast machine and believed in the theoretical possibility of determining all past and future events from the present state of the world, appears to us today as hopelessly naive. This picture won't even do as a thought experiment, since according to our current understanding of the world on the smallest scale, the exact measurement of one quantity is always linked to a random change in another quantity over which one has no influence.

However, sometimes, this "sensitive dependence on the initial state" does not occur: the motions of the planets can be predicted to extreme accuracy long into the future. With the weather, on the other hand, science quickly arrives at the limit of its ability to predict. Whether the wedding you have planned for next June should be held outdoors or indoors will have to be decided at the last minute. And that situation is not about to change: one never knows when a butterfly will get it into its head to flap its wings.

Linearity versus Nonlinearity

The topic of chaos theory provides us with the opportunity to say something about the term "linear," which appears in a number of fields with various meanings. If one is speaking about a computer program, then "linear" means that the various instructions are to be carried out in sequence, one at a time. As an alternative, one has parallel processing, whereby dozens or thousands of processors work together, with many instructions being executed simultaneously.

And until a few decades ago, it was usual to take in information, such as that in a book like this one, "linearly": one reads line by line, starting at the beginning and finishing up at the end. But that is now terribly passé, for now one can obtain information in quite another way. Anyone surfing the Internet can click on a word that is highlighted as a link, go to another web site altogether to read about that concept, and then return to the original text or else continue on his or her travels through the worldwide web. This mode of acquisition of information seems to correspond particularly well to the way that we think.

In mathematics and physics there is another, more narrowly defined meaning of the word: a process is called "linear" if a superposition of input values leads to a superposition of the corresponding output values. For example, if a system returns output F on input f, and output G on input g, then an input of $f + g$ will lead to an output of $F + G$. A simple example is the (not too large) stretching of a spring. If the spring extends by 5 inches when a force of 3 pounds is applied, then a force of 6 pounds should result in an extension of 10 inches. The following facts play an important role in the natural sciences:

- On a small scale, many physical processes are approximately linear. That is because most processes in nature are relatively uniform, without pathological jumps, and therefore the graphs that represent them are rather smooth. Since a small piece of a smooth curve can be fairly well approximated by a straight line (the tangent), on a small scale the process appears to be more or less linear.

- On the other hand, in nature there are no processes that are linear in a strict sense. If one stretches a spring beyond a certain point, the deformation of the spring's structure and stresses in the metal will cause the process to be no longer linear, and if you stretch the spring to the breaking point, you can kiss linearity good-bye!

- The linear approximation of a system leads to considerable simplification in analysis. That is because one can concentrate on particularly simple solutions, with additional solutions being inferred by superposition. For example, the sound produced by a trumpet is composed of simple vibrations, namely the fundamental frequency and the harmonics above it. One can hear the higher frequencies by changing the frequency of the embouchure: the trumpet will "speak" at the fundamental tone, the octave, a fifth above that, two octaves above the fundamental, and so on.

Therefore, "nonlinear (fill in the blank)" is in principle more complicated than "linear (fill in the blank)," where contemporary mathematics offers countless examples for filling in the blank: nonlinear operators, nonlinear partial differential equations, and so on. It is clear that most of the interesting problems in the real world, such as the weather, chemical reactions, and the development of the cosmos, lead to nonlinear problems. And it is just such problems that force us to deal with truly chaotic behavior.

Chapter 83

Guaranteed to Make You Rich

Have you ever had a prophetic dream? You dream that your Aunt Maude has telephoned you, and the next evening, the phone rings, and there she is. Are such phenomena explicable by ordinary processes, or are there higher powers at work? Such occurrences in fact do not topple the foundations of modern science, for the explanation is rather simple: when an experiment with even a small probability of success is performed a large number of times, one can expect the experiment sometimes to succeed.

As an illustration of this assertion, imagine a large room filled with people. Each of the many persons present is asked to think of a number between one and six. And now a die is rolled. Regardless of the number that comes up, about one-sixth of those present—namely, those who chose that particular number—might well have the feeling of having accurately predicted the future. And as for Aunt Maude's telephone call, something similar is at work. With so many dreamers dreaming so many dreams, it is unavoidable that occasionally dream and reality should coincide.

Thus it is that one sometimes reads something in one's horoscope that actually "predicts" something that then actually occurs. When a large enough number of people read a prediction (particularly when it is rather vague and could apply to almost anyone), it will certainly be true for some of them.

We now present a theoretical practical application of this phenomenon. (The author abjures all moral and legal responsibility for anything resulting in someone's actually trying this out.) Send a postcard to each of one thousand individuals who have an interest in horseracing on which you have written a prediction as to the outcome of a particular race. Say there are ten horses scheduled to run this race; then you are to predict each of these horses on one hundred of the cards. Whatever the outcome, one hundred recipients of your card will have received an accurate prediction. To these one hundred you send predictions for the next race, with the first ten persons on your list being given the prediction that horse number 1 is going to win, the next ten, horse number 2, and so on. After this race, ten of the recipients will have received yet another accurate prediction. And now for the third round: you send ten cards to these ten persons, and one of them will again have a winner and believe that you can see into the future.

Ask this person how much he is willing to pay for a tip on the next race. Certainly it will be more than you have spent on postage.

The moral is this: whoever makes a large number of guesses will almost certainly be right some of the time. And among the thousands of Aunt Maudes out there, why shouldn't one or two decide to telephone their nephews this evening?

The Pole on the Highway

The phenomenon that underlies all that we have been discussing in this chapter is a further example of the fact that we are unable to grasp very large numbers. How else can we explain that even though the chances of winning the lottery's grand prize are depressingly small (1 in 13,983,816, or about one in fourteen million), nevertheless, almost every weekend yet another winner is announced?

To make this clearer, let us consider a further illustration: Imagine a stretch of highway 220 miles long, roughly the distance between New York and Boston. Now, 220 miles is equal to $220 \cdot 12 \cdot 5{,}280 = 13{,}939{,}200$ inches. The probability of a person guessing the particular inch of highway that you are thinking of is just about the same as the probability of winning the lottery. Suppose your assistant erects a pole one inch in diameter somewhere on the highway between New York and Boston (see Figure 1), and you—as a passenger, of course—ride blindfolded between the cities. Do you think you will hit the pole if at some point during the trip you toss a penny out the window? This seems unlikely in the extreme.

Figure 1. Can you hit the pole?

Every weekend, millions put down their money on the lottery. In our highway analogy, this means that over the course of many weeks, the highway between New York and Boston is packed with bumper-to-bumper traffic,[1] each car with a passenger tossing one penny out the window at some point on the trip. It is certainly not unlikely that one of the coins in this shower of copper will hit the pole.

[1]Twenty million bets and an average automobile length of 15 feet makes about 57,000 miles, or more than two times around the Earth.

Chapter 84

Don't Trust Anyone over Thirty

One is always hearing that the greatest accomplishments in the history of mathematics were made by very young researchers. Is that true?

It is true that through the centuries, mathematics has been moved in significant directions through the ideas of youngsters who today would be high-school or college students. Evariste Galois (1811–1832, see the figure) died in a duel at the tender age of twenty, not long after making a discovery that would revolutionize the field of algebra: how one can tell from a polynomial equation whether it is solvable using the usual operations of addition, multiplication, and the extraction of roots. Or consider Niels Henrik Abel, about whom we had much to say in Chapter 72. This Norwegian mathematician lived only to age 26. A letter addressed to Abel arrived two days after his death from consumption informing him that he had been offered a professorship at the University of Berlin. Abel is considered Norway's greatest mathematician by far, and he was honored, much belatedly, a few years ago with an endowed prize in his name of almost one million euros. This Abel Prize was conceived as the mathematical equivalent of the Nobel Prize.

There are examples to be found in our times as well, and at every major conference, astonishingly young speakers present extremely mature results. The most prestigious of the mathematics prizes, the Fields Medal, is directed at such individuals. One can receive a Fields Medal only if one has not attained one's fortieth birthday when the prize is awarded. These prizewinners are sought after for the world's best-endowed chairs of mathematics.

The prize committee had to contort itself considerably in order to honor the mathematician Andrew Wiles at the International Congress of Mathematicians in Berlin in 1998. By general consent he had made the most important contribution to mathematics of the past century—the proof of the Fermat conjecture—but he was well past forty.

We should therefore not take too seriously the thesis that in mathematics as in athletics, one is all washed up by the mid thirties. There are many famous mathematicians who continued their creative output throughout a long life, where surely the best-known name in this regard is that of Carl Friedrich Gauss (see the figure), who lived from 1777 to 1855.

Perhaps one would do better to compare mathematicians with orchestra conductors: working with fascinating material keeps the gray cells in shape well into old age.

Chapter 85

Equality in Mathematics

What is essential in a mathematical problem and what is really just window dressing? To be more precise, we would like to understand when two situations are "equal," so that we can limit our effort to mastering only one of them in order to understand both.

What, then, does equality mean in mathematics? Surprisingly, it is similar to the situation in our everyday lives, where equality in the sense of being identical plays far less of a role than equivalence with respect to some aspect.

For jotting a quick note, a napkin can be as good as a pad of paper. Or with respect to the goal of getting to the opera this evening, a compact car, a limousine, and a taxi are all equivalent. However, as soon as cost, estimated time of travel, or prestige comes in the door, any idea of equivalence flies out the window.

The situation is similar even in elementary mathematics. If you want to explain to your child the meaning of "five," you can as well produce five apples as five children. The apples and the children, with respect to "fiveness," are entirely equivalent.

The complete truth is somewhat more complicated. If one wishes to explain what the abstraction "five" really means, one generally begins nowadays with "equivalent with respect to number." Then "fiveness" is the entirety of all objects that are equivalent in respect to the set of the fingers on one hand.

This principle permeates all of mathematics. In geometry, two triangles are equal if one can be superimposed on the other by a translation, rotation, and possible flip. And in probability theory, it makes absolutely no difference whether one uses a fair coin or a fair die in making a decision based on chance.

(If the decision is based on the coin's coming up heads or tails, the decision with the die could be used based on an odd or even number, for example.)

Only through the notion of equivalence can one bring order into the immeasurable richness of mathematical objects. The same principle obtains in every language, where certain general concepts make communication possible. Even if we don't have exactly the same notion of "flower" or "beautiful," language is still able to function quite well with such notions of equivalence.

Chapter 86

Magical Invariants

What remains unchanging? On what can one rely? For centuries, mathematicians have searched for invariants, that is, to put it simply, quantities whose values do not change with respect to some operation under consideration.

As an example, let us consider a deck of cards. When you shuffle the deck, there are certain invariants in play. Certainly, the number of cards is unchanged, as are the numbers of jacks, queens, and so on. The situation is different when one does not allow shuffling, but only repeated cutting of the pack and putting the bottom "half" on top of the top "half." Then the relative order of the cards remains unchanged. If the ace of spaces was to be found three cards down from the queen of hearts, it will remain so.

Of course, we have to interpret the words "down from": If the queen of hearts happens to be the bottom card in the deck, then the ace of spades will be three cards down from the top. That is, "down from" is to be understood in the sense that when you reach the bottom of the deck, you continue from the top.

One can make use of this invariant in a little magic trick. Remove the kings and queens from the pack and lay them out in a row, as shown in Figure 1. The key here is to make the distance between cards of the same suit (king of spades and queen of spades, king of clubs and queen of clubs, and so on) exactly four. If you flash this arranged hand of cards in front of your audience, it appears to be more or less randomly arranged. No one will suspect your skulduggery, and if you cut the pack a few times (Figure 2), everyone will believe that the cards have been thoroughly mixed.

Figure 1. The prepared cards.

But you are operating with the knowledge that the distance between cards is an invariant: four cards below the first card is its partner. It is therefore easy to produce a pair from under a cloth or under the table, of course making it seem that you are struggling mightily. You can repeat this process and produce a second pair (though this time the distance between them is three), and then a third pair (separated by two), and finally the last pair.

This trick relies on the existence of some kind of order amidst seeming chaos. In mathematics, the search for the unchanging has become a sort of leitmotiv in research. Once a set of permissible transformations has been described, a systematic search begins for quantities that remain unchanged under those transformations. This idea has been of particular importance as a unifying principle in many branches of geometry. It was proposed in 1872 by the mathematician Felix Klein and has had great influence over research ever since.

Figure 2. The cards after being cut.

The Back Story: The Distance Modulo the Number of Cards Is Invariant

Using the notion of modular arithmetic introduced in Chapter 22, we can formulate the principle that underlies the trick somewhat more mathematically:

> If n cards are stacked, and two of these cards are in positions a and b (counted from the top), then $(b - a)$ modulo n is an invariant: after cutting the pack any number of times, the difference between the positional values has not changed modulo n.
>
> To see this, one has to use modular calculations for negative numbers as well as positive, but that is really no problem. After all, as everyone knows, the day of the week seven days *ago* is the same as the day today, and the day 13 days ago was Tuesday if today is Monday. Mathematically, -13 modulo 7 is equal to 1.
>
> One must be aware of this subtlety in order to interpret the invariant relationship presented above correctly. An example: In the example that follows we are going to use the fact that -7 modulo 10 is equal to 3. In a pack of ten cards, the ace of hearts and jack of clubs are in positions 2 and 5. The difference is 3. The pack is cut at position 2. Now the ace of hearts is at position 10, the bottom of the pack, and the jack has moved up to position 3. The difference (position of the second card minus that of the first) is thus $3 - 10 = -7$, and modulo 10, this is the same number, 3, as before.

We Draw on an Extensible Surface

Only a few mathematical invariants are suitable for magic tricks. Their great significance is that invariants separate the essential in a theory from the inessential. To show this via a somewhat unconventional example, we require a drawing surface made out of some stretchable material.[1]

On our surface we draw a figure: a triangle, a circle, a collection of rectangles, whatever. Now the surface is distorted; we pull it and compress it in any way that strikes our fancy. Our drawing will change significantly. A small circle can become a large circle; a right angle can become obtuse or acute.

If the original figure had the property that one could connect any two points with a curve contained entirely within the figure, which is the case for a circle or triangle, but not a collection of rectangles, then one can still do so after the figure has been altered: *connectivity* is an invariant under distortion.

[1]Perhaps a piece of one of those elastic exercise bands.

Chapter 87

Mathematics Goes to the Movies

Every now and then, mathematics makes an appearance at the cinema. Mathematicians go to such films with a certain degree of trepidation, for often there is nothing more than the latest clichés to be seen. Nevertheless, it is always interesting to see what aspect of the subject the writers and directors have chosen to display.

Consider the 1992 film *Sneakers*. The good guys, led by Robert Redford, attempt to wrest from the bad guy (Ben Kingsley) a device developed by a brilliant mathematician for cracking all the world's secret codes.

Before suffering a violent death, the mathematician is seen at a conference. What he says there is certainly something that well could be heard at such a gathering. It would appear that for a change, the writers and director did some serious research. The people actually speak in such a way that anyone with a semester of mathematics could understand everything they are saying. Mathematics comes out rather well in the film, even though the abilities of mathematicians to decode all the world's encrypted secrets are greatly exaggerated.

Exaggeration heads in another direction in the film called simply "π," where mathematical mysticism goes off the deep end. The idea is that many secrets are encoded in the decimal digits of π, and if one reads them correctly, many strange phenomena suddenly seem to be explained. Perhaps it should be taken metaphorically: indeed, π plays an important role in just about every area of mathematics and there are many mysteries about the number that are waiting to be discovered.

If a poll were taken among mathematicians to discover their favorite film with a mathematical theme, the winner would certainly be *A Beautiful Mind*, with Russell Crowe in the leading role. The movie is an adaptation of Sylvia Nasar's biography of the game theorist John Forbes Nash. The emotional aspects of mathematics are captured in the film remarkably well. The irresistible compulsion to solve a problem can become so all-consuming that one's personal life is endangered.

The moral is this: anyone considering marrying a mathematician should be prepared to live with someone who frequently disappears into another world. It is the rare mathematician who can simply call it a day and put off wrestling with a problem until tomorrow.

Chapter 88

The Lazy Eight: Infinity

Mathematicians have to deal with the infinite on a daily basis. It appears in a variety of guises. The most harmless variant is the infinity that arises in counting. One begins with 1, comes then to 2, proceeds to 3, and so it goes. An end is never reached. Even the most critical researcher on the foundations of mathematics would agree that this infinity is unproblematic.

Things get more awkward when infinite quantities are treated as new mathematical objects. Can one really speak about the set of all prime numbers? Even if no one knows how to determine whether a particular very large number is prime? It is now generally agreed that it is legitimate to speak of such a set, and the number of critics continues to shrink.

For mathematicians interested primarily in applications, such basic questions are of secondary interest. For them, "infinite" means simply that some quantity is incommensurably larger than some other quantity. The mass of the Sun can in some situations be considered as infinitely large in comparison to the mass of the Moon. Bill Gates's fortune can be considered infinitely large in relation to your bank balance, and so on.

Eventually, one becomes accustomed to working with infinite quantities, and one can calculate with infinity as easily as with finite numbers. For example, one has the rule that "infinite plus finite

Figure 1. Thus was the infinite universe conceived in the Middle Ages.

equals infinite," which is reflected in the fact that Bill Gates doesn't really become any wealthier if you give him everything you have. Or that a battleship becomes no heavier when a flea alights on deck.

Using this idea, many computations can be simplified. For example, if one wishes to know how the three celestial bodies Moon, Sun, Earth interact, it is a great simplification to consider the mass of the Sun to be infinite.

None of this is anything new. Almost five hundred years ago, Copernicus considered a problem that could be solved only by bringing in the idea of infinity. How could one explain that the positions of the stars remain seemingly unchanged during the revolution of Earth about the Sun? Copernicus solved the problem with remarkable elegance: the distance from Earth to the nearest star is infinite in comparison with the diameter of Earth's orbit. That explained the phenomenon, though it invited in a host of theological problems (see Figure 1). Suddenly there was no place in the universe for God as then conceived, and it took the Church several centuries to accept the Copernican model of the solar system.

How to Calculate with "∞"

Mathematicians use the symbol ∞, a "lazy eight," to represent infinity. It can be imagined as a special kind of number. If one represents the "usual" numbers by a line stretching without end to the left (negative numbers) and to the right (positive), then ∞ would be represented by a point to the right of the line (and $-\infty$ would be a point to the left). This expresses the fact that ∞ is greater than every "normal" number.

One would like to extend the usual mathematical operations, as far as possible, to the infinite realm. We have already mentioned that addition is defined in such a way that the sum of any number and infinity is again infinity.[1] The product of infinity and a positive number is also infinity. This can be written as $a \cdot \infty = \infty$ (for positive a). This is again plausible, since Bill Gates would remain immeasurably wealthy even if some foolish financial transaction caused his fortune to be reduced to half its value.

However, caution is advised, for one must forswear some of the rules to which one has become accustomed. As an example, consider the rule for "normal" numbers that allows one to conclude from $a + x = b + x$ that $a = b$. (This is an abstract representation of a rather commonsense notion: if Mr. A and Ms. B celebrate their fortieth birthdays on the same day, then they must have been born on the same day forty years ago.)

But once addition of infinities is allowed, this rule has to yield: for example, $10 + \infty = 1{,}000 + \infty$ (both sums are ∞), but one may certainly not conclude from this that $10 = 1{,}000$.

[1] This can be written as $a + \infty = \infty$.

Chapter 89

Books Need Bigger Margins!

It has been mentioned more than once in this book that mathematics is not always directed toward useful applications. Even when no such application is on the horizon, thinkers can be motivated to unbelievable intellectual accomplishments if a problem is sufficiently compelling.

A famous example is the solution of the Fermat problem. Almost 400 years ago, in 1621, the French mathematician Bachet translated Diophantus's treatise *Arithmetica* from Greek into Latin. It was read by the jurist and amateur mathematician Pierre de Fermat (1601–1665), shown in Figure 1, who was inspired by Diophantus to grapple with the question of higher-dimensional variants of *Pythagorean triples*, integers a, b, c such that the sum of the squares of the first two equals the square of the third ($a^2 + b^2 = c^2$). There is an infinite supply of such triples, of which $3, 4, 5$ is the best known (observe that $3^2 + 4^2 = 9 + 16 = 25 = 5^2$). A triangle with sides satisfying this condition is of necessity a right triangle, and this fact can be used to good effect in the construction of right angles, for example in constructing a garden.

Fermat wondered what would happen if one replaced the word "squared" by "cubed" (third power) or some higher power. For example, are there whole numbers a, b, c that satisfy the condition $a^4 + b^4 = c^4$? Fermat became convinced that other than trivial solutions with a or b equal to zero, no such numbers could exist, and he was able to prove this for fourth powers. Apparently, he believed that his method of proof for exponent 4 would carry over to other exponents as well, for in the margin of his copy of Diophantus, he wrote (in Latin), "I have discovered a truly marvelous proof of this, which this margin is too narrow to contain."

For over three hundred years, mathematicians by the score (and amateurs by the hundred) attempted to prove Fermat's conjecture or to find a counterexample. It became perhaps the most famous problem in the history of mathematics. The massive effort at finding a proof was motivated in part by a sort of athletic ambition—to triumph on the intellectual playing field: "So many have tried and failed. If I were to solve it...." In another vein, the intrepid search—long in vain—for a solution led to enormous strides in our understanding of algebra.

Figure 1. Pierre de Fermat and Andrew Wiles.

All the world now knows that Fermat was correct.[1] In 1998, the English mathematician Andrew Wiles (see Figure 1) completed a proof on which he had worked almost his entire academic life. Alas, we never shall know for sure whether Fermat actually found a proof.

[1]Not that he had found a proof, but that there are no higher-dimensional analogues of Pythagorean triples.

However, the methods developed by Wiles and others that were necessary for Wiles's proof are so deep and require so much machinery of modern mathematics developed since Fermat's time that it is as good as impossible that Fermat had a valid proof of his conjecture.

The Method of Infinite Descent

The Fermat problem is a good example of how different the difficulties can be between proving that something can be done and proving that something cannot. Let us illustrate this for the exponent 4. Suppose that it were in fact the case that there existed three natural numbers a, b, c such that $a^4 + b^4 = c^4$. Then one could write a computer program and hope that such numbers would eventually be found. If after a year, the computer had failed to find such a triple, one would begin to wonder: it could be that the only such triples are numbers astronomically large, in which case the computer is going to be of no help.

But what if you have the suspicion that there is *no solution*? Not even with hundred-digit numbers, nor with numbers so large that to write them down would consume all the ink that has ever been produced? Now we are faced with a much more difficult problem. The strategy is similar to the proof of the irrationality of the square root of two (see Chapter 56): assume that the assertion is false (that is, assume that the square root of two is rational), and then provide a deductive argument that leads to a contradiction. Therefore, the assertion that you are trying to prove must be correct.

This idea can be used in a somewhat different form to prove the Fermat conjecture for the special case of exponent 4. Once one has mastered some basic results of number theory, the proof can fit on a single sheet of paper. The magical incantation is "infinite descent," and it involves the following method of proof.

One shows that *if* there existed natural numbers a, b, c such that $a^4 + b^4 = c^4$, then one could also find natural numbers d, e, f with the same property, namely $d^4 + e^4 = f^4$, and with the additional property that f is smaller than c. In other words, for every triple satisfying the Fermat equation with exponent 4, there is another triple that satisfies the equation with a smaller number on the right-hand side. But that is impossible, for it implies an infinitely descending sequence of natural numbers, and since these numbers stop at 1, no matter how large a number c you start with, there are at most c numbers in a descending sequence. For instance, a sequence of natural numbers beginning with 5 can have at most the members $5, 4, 3, 2, 1$, while if you start with 100,000, you can go on for a while longer, but eventually you will get down to 1, and the game is over.

Unfortunately, the method of infinite descent works for only a few exponents in the Fermat problem. Wiles's proof relies on much deeper results and methods, and indeed there is only a handful of mathematicians who could justifiably claim that they understand every detail.

Chapter 90

Visualizing Internal Organs with Mathematics

That a mathematician is sometimes a detective is something that one learns early on at school. For example, x can be an unknown quantity about which one knows only that $3x+5=26$. Sherlock Holmes, take over! If $3x+5=26$, then we must have $3x=21$, and then x is revealed as being equal to 7.

In computer-aided tomography (CAT), there are analogous problems, though on a much higher level. As an illustration, imagine a plane figure, say a circle, an ellipse, or a rectangle. We call in a glazier, who cuts us a copy of this figure out of half-inch-thick glass.

Now hold the glass figure with its thin edge pointing toward the light. If our figure was a circle, the light will have a much shorter path to travel at the top and bottom than in the middle, where it travels through the entire diameter of the circle. Thus the middle will appear dark—likely a dark green—and the edges will be much lighter. On the other hand, if we view a rectangle from the side, we will see a stripe of uniform brightness.

And now the \$64,000 question: by measuring the differences in brightness from a variety of directions, can one determine what shape is present? Surprisingly, the answer is yes indeed, and that is the basis of computer-aided tomography (a computer tomograph is shown in the picture). The problem in this technique of medical diagnosis is quite similar to our glass problem: the human body is x-rayed from a number of directions, measurements are taken of the intensity of absorption, and these values are used to compute a three-dimensional image of a specific organ of medical interest.

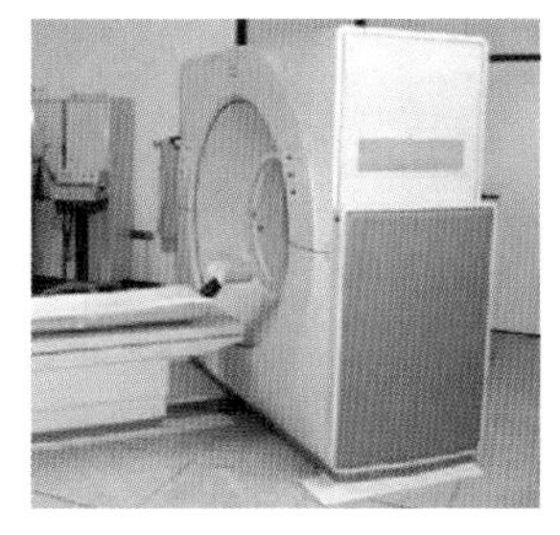

That's the general idea; the details are painfully complex. It took a combination of engineering skill, computer technology, and rather advanced mathematics to create what today is a standard tool of medical practice.

It took only a few years from the concepts developed in the 1960s to a practical implementation. One reason for such a speedy result is that most of the mathematics had already been developed and was just sitting on the intellectual shelf, waiting to be exploited. Almost one hundred years ago, the mathematician Johann Radon (1887–1956) proposed a procedure for reconstructing illuminated objects from measurements of intensity.

A computer tomograph is thus not merely "high tech," but also "high math," so to speak. Research in this area continues, since there is much room for improvement in both speed and resolution.

Inverse Problems

The problem that arises in computer-aided tomography is a special case of an *inverse problem*. Such problems arise in a number of practical applications. For example, when vibrations from under the Earth's surface are measured from a number of different locations, it is possible to determine exactly where an earthquake occurred and with what intensity. Similarly, measurements of reflected waves can provide information on the location and amount of various types of subterranean mineral deposits.

All inverse problems exhibit certain typical difficulties, which therefore also play a role in computer-aided tomography. For example, there is the sensitive dependence of the solution on the measured values: slight errors in measurement—and no measurement can be one hundred percent exact—can lead to large errors in the reconstruction.

For an illustration of such difficulties, consider the equation $0.0001 \cdot x = a$. The number a is a known quantity, and x is a large unknown number. Mathematically, the solution is easy: $x = a/0.0001 = 10{,}000a$. However, in a real-world application, the value of a will not be known exactly. If a represents a length, its value might be known to within one-tenth of an inch. But the possible error in the calculation of x is then ten thousand times as big, and so the computed value is accurate only to within over eighty feet!

Chapter 91

A Brain in the Computer

The mathematician as Frankenstein? For centuries, people have cherished hopes of transferring some of mankind's intellectual capabilities to machines. Neural networks, which have been an object of study since the 1960s, represent a serious attempt to simulate aspects of the brain's structure and function in a computer with the goal of creating something approximating human thought.

The fundamental building blocks of the brain are neurons: impulse-conducting cells that form the central command and control structure of the nervous system. Human beings have about ten billion of these cells, wired together with several trillion connections, or *synapses*. The computer equivalent of a neuron is a logical structure that either strengthens or weakens a signal received as input based on the state of certain control signals. The response can vary widely, depending on the reaction to the control signals, and by setting various parameters, one can achieve an enormous range of possible behaviors. By linking several of these structures together, the number of variations becomes exponentially vast, and one speaks of such a linked collection as a *neural network*.

How does one begin to choose the various parameters? As a simple example, let us consider the problem of how a bank might determine whether to approve your application for a credit card using the information that it is able to obtain about you: age, income, assets, credit history, and so on. The ideal would be a neural network that took in all of this information and output either yes or no in such a way that it is precisely the creditworthy applicants who are approved.

After constructing a network, one "trains" it by using as input a number of cases in which it has already been determined by other means whether the applicant should be offered a credit card. The idea is to tweak the parameters in such a way that the correct answer is given for all the members of the training set. Here one requires some fairly sophisticated mathematics. The hope and expectation is that with sufficient training, the neural network will give the correct answer in situations that are new both for the computer and for the bank.

Classical mathematics takes a rather skeptical attitude toward such methods, for the process of manipulating the model to correspond to reality disregards any actual understanding of the interrelationships among the various parameters. However, one may well ask whether the "gut" determinations of a bank employee are any more reliable than those of a well-trained neural network.

The Perceptron

How, then, are brain cells simulated in a computer? One of the earliest proposals was the *perceptron*, which was studied in the 1960s. In its simplest manifestation one can imagine it as a black box with a number of input wires entering at one end and a single output wire emerging at the other (Figure 1).

What the perceptron does is to multiply the input signals $x_1, x_2, \ldots$ by a set of weight factors $w_1, w_2, \ldots$ and then add the products. Then it is determined whether the sum $w_1x_1 + w_2x_2 + \cdots$ exceeds some threshold value T. If that is the case, the output voltage is set to 1, and one says that the perceptron "fires." If the threshold value is not exceeded, the output is set to zero.

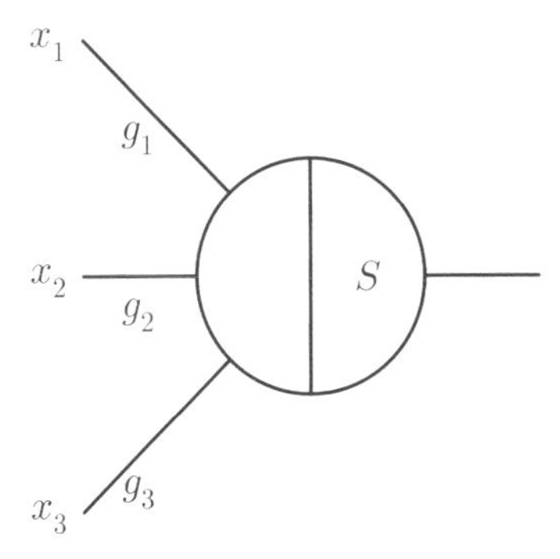

Figure 1. A perceptron.

As an example, consider the case in which there are two input signals, the threshold value T is set to 1, and the two weights have the same value of 0.7. If now a voltage of 1 is applied to one of the inputs, while the voltage of the other input remains at zero, the sum of the products of weights and voltages is equal to $0.7 \cdot 1 + 0.7 \cdot 0 = 0.7$, which is less than the threshold value. Therefore, the output remains at zero, and the perceptron does not fire. However, if both inputs have voltage 1 applied, then the sum attains the value 1.4, and the perceptron fires.

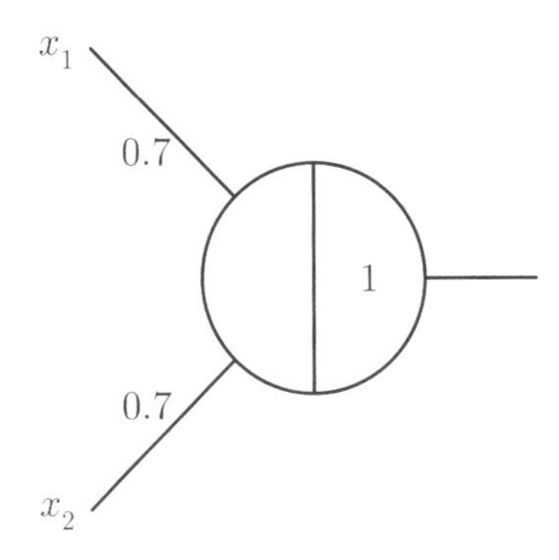

Figure 2. The perceptron as an AND gate.

One way of looking at the situation just described is that with a suitable choice of weights and threshold value, a perceptron can realize an AND gate (see Figure 2).

But the perceptron can do much more. Many readers will recall from their schooldays that the set of all points with coordinates (x, y) satisfying an equation of the form $ax + by = c$ constitutes a straight line in the two-dimensional Cartesian plane. Those pairs (x, y) for which the sum $ax + by$ is greater than c are precisely the points lying to one side of the line, as shown in Figure 3.

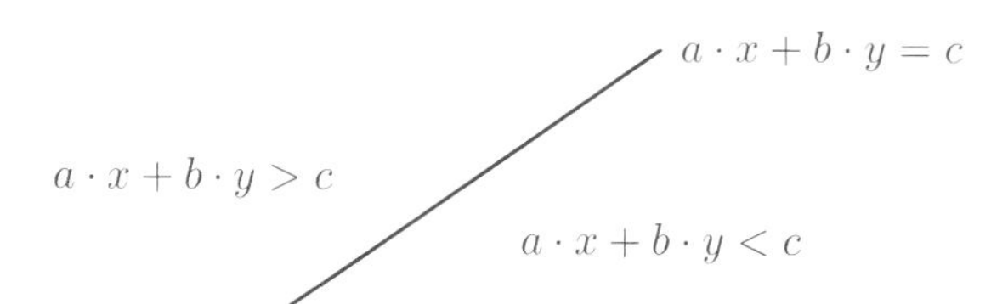

Figure 3. A half-plane is determined by the inequality $ax + by < c$.

Returning to the perceptron, we let x, y be the input voltages, a, b the weights, and c the threshold value. Then the perceptron can determine whether a point lies on a particular side of the line. Since AND gates can be linked, one can wire a number of perceptrons together to model a miniature brain that outputs 1 if a point lies inside a given triangle and 0 otherwise. With the coordinates of the points as the input voltages, the implementation is simple, as can be seen in Figure 4. A point lies inside triangle ABC if it lies to the right of line $G1$ AND it lies to the left of line $G2$ AND it lies above the line $G3$.

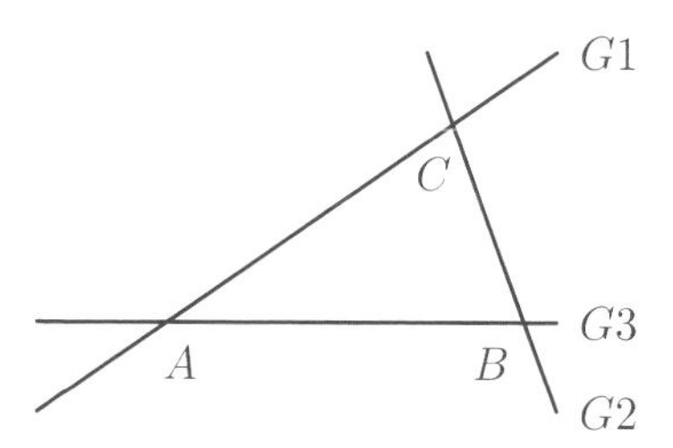

Figure 4. A network of perceptrons can determine whether a point lies inside a triangle.

In more-serious applications, several dozen perceptrons are linked, and again one speaks of a neural network. The choice of weights for the input signals is determined by a trial-and-error approach: better choices of weights are those that produce an output closer to what is known to be correct. In the credit-card example, if the input values for income, home ownership, length of time in current job, etc., are favorable, the neural network should output 1 (creditworthy!), while candidates with values that suggest a credit risk should be rejected.

Chapter 92

Cogito, Ergo Sum

René Descartes (1596–1650) was a remarkable personality (see Figure 1). He decided at a young age to devote his life to the pursuit of knowledge. In his *Discourse on Method* of 1637 are to be found not only the basis of his philosophy ("cogito, ergo sum"), but three important appendices in which he attempted to demonstrate his new method.

One of these appendices is devoted to geometry. In it there appear several proposals that were to have a profound effect on the development of mathematics. The most important of these was surely the unification of algebra and geometry. Descartes recognized that geometric problems could be translated into the language of algebraic equations and that the solution of equations had in many cases a geometric interpretation. This approach turned out to be enormously fruitful, for problems that could not be solved in one domain could frequently be solved in the other. A particularly spectacular example was the proof of the impossibility of squaring the circle:[1] since the number π was eventually shown to be a particularly complicated type of number,[2] and only relatively simple numbers can be constructed with straightedge and compass, it is impossible to transform a circle into a square of the same area using such a geometric construction.

[1] See Chapter 33.

[2] The number π is *transcendental*. See Chapter 48.

Figure 1. René Descartes (1596–1650).

However, for Descartes, all of that lay far in the future. Too little was known about numbers. Even negative numbers were eyed with suspicion, and in working with equations, one had to deal with the cumbersome notion of "true" and "false" solutions depending on whether they were positive or negative.

The rapid development of the natural sciences in the seventeenth century is unthinkable without the preparatory work done by Descartes. It is difficult to believe that he was "only" an amateur mathematician and an autodidact in the subject.

We should mention that the "Cartesian coordinate system" that we all learned about in school is nowhere to be found in Descartes's work. It was only in the eighteenth century that it was recognized that such a system could unify the various geometric constructions that had theretofore been treated as special cases.

The Pythagorean Theorem Is "Translated"

As an example of the power of the translation from geometry to algebra, we shall demonstrate how easily the Pythagorean theorem can be proved when it is translated from geometric figures to algebraic equations.

What has to be shown is that in a right triangle with sides a, b, and c, the equation $a^2 + b^2 = c^2$ holds. We shall assume that if a and b are unequal, then b is greater than a. See Figure 2.

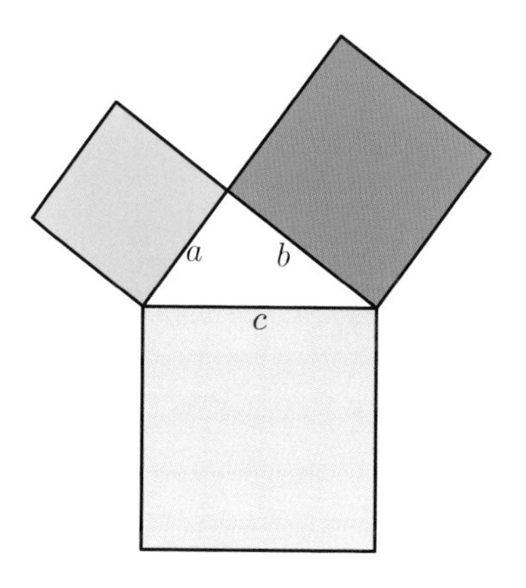

Figure 2. The Pythagorean theorem: $a^2 + b^2 = c^2$.

The secret to the proof is to consider a square with side length c with four copies of our triangle inscribed within.

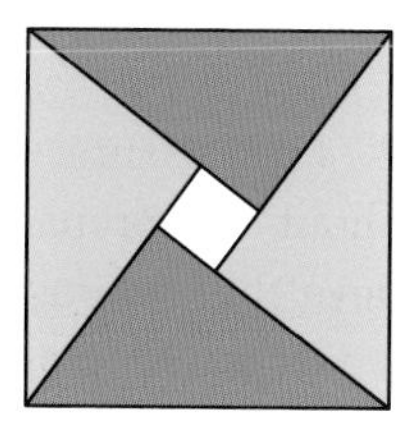

Figure 3. The Pythagorean theorem: A proof.

The four triangles do not fill up the square: in the middle there remains a small square (the white square in Figure 3). From the figure it is clear that this little square has side length $b-a$. From this we conclude that the area of the big square is equal to four times the area of the triangle plus that of a square with side length $b-a$. Since a right triangle with legs of lengths a and b has area $\frac{1}{2}a \cdot b$, we have the equation

$$c^2 = 4 \cdot \frac{a \cdot b}{2} + (b-a)^2.$$

Now we engage in algebraic legerdemain. Recalling that $(a-b)^2 = a^2 - 2 \cdot a \cdot b + b^2$, we have proved that

$$c^2 = 4 \cdot \frac{a \cdot b}{2} + (b-a)^2 = 2 \cdot a \cdot b + b^2 - 2 \cdot a \cdot b + a^2 = a^2 + b^2.$$

We have thus shown that $c^2 = a^2 + b^2$ by replacing a complicated geometric argument with a simple algebraic one.

This example should not leave the impression that Descartes was concerned solely with such comparatively simple problems. On the contrary, most of the questions that he considered are so deep that many mathematicians today—almost four hundred years later—would have difficulty finding a solution.

Chapter 93

Does the World Have a Hole?

The Clay Mathematics Institute has offered million-dollar prizes for a number of problems in mathematics that have long defied solution.[1] The consensus among mathematicians is that the Poincaré problem will be the first to fall. That would be a sensation indeed, since generations of mathematicians have been gnashing their teeth over their failure to prove Poincaré's famous conjecture.

The Poincaré problem involves an understanding of "space." Before we move into three dimensions, let us consider the problem in two dimensions, where visualization is easier. What are the categories of surfaces that are "essentially" different? The agreed-upon definition of essential difference is that two such surfaces are essentially different if one cannot be continuously deformed into the other.

In this sense, the surface of an orange is essentially the same as that of the Earth or of a football. However, the surface of a life preserver is something different altogether. See Figure 1. In the nineteenth century, work began on a complete catalog of surfaces, and since then, the classification of surfaces has been successfully completed.

[1]See Chapters 57 and 37 .

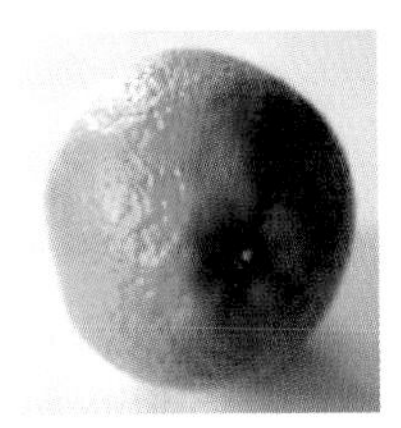
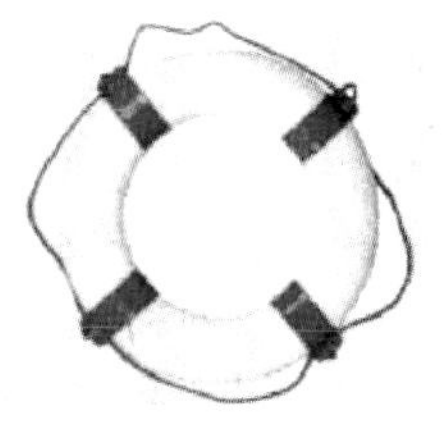

Figure 1. Two essentially different surfaces.

For the corresponding question in three dimensions, that is, for three-dimensional space, the problem seemed hopeless. In order to explain the Poincaré problem, we need to introduce a bit of terminology, namely the notion of simple connectedness. Imagine your living quarters with all the furnishings removed, the front door closed, and all the internal doors removed from their hinges. Now take a long thread, pull it around through the various rooms in any way that you like, and then tie the two ends together. If you pull on the loop, either the whole thread will glide into your hand, or else it will get hung up somewhere because your dwelling allows for a circular path through the rooms, as in the right-hand picture in Figure 2.

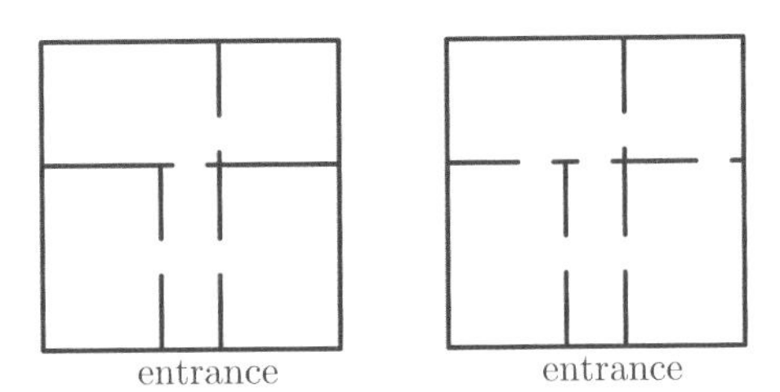

Figure 2. Is your home simply connected?

So here's the definition: A region of space is said to be *simply connected* if, as in the left-hand picture of Figure 2, it is impossible for a thread as described above to get hung up. (This definition can also be applied to two-dimensional surfaces. The surface of a sphere is certainly simply connected, while the surface of a life preserver is not.) Poincaré conjectured that there is only one region in space that is simply connected and in a certain technical sense is "not too large."

That was around the year 1900. Since then, enormous progress has been made in our understanding of space, yet the problem remained unsolved. This situation was most unsatisfactory, since in the meanwhile, problems in other areas of mathematics of apparently much greater difficulty had been solved. However, it seems now that a solution is imminent. This cautious formulation is necessary, because the proof strategy proposed by the Russian mathematician Grigoriĭ Perelman has not yet been worked through in full detail.[2] However, mathematicians are fully convinced of the validity of an argument only after it has been carefully expounded with complete rigor.[3] And that could take quite a while, though this time, the profession is highly optimistic about the outcome.

The wait will be worthwhile, for if Perelman's ideas bear fruit, much more is at stake than the original problem. We would then have a complete catalog of the architecture of all possible spaces, and one could construct any of them out of an erector set with eight different types of building block. Such was predicted by the American mathematician William Thurston in the 1970s, but no progress toward the realization of the Thurston program was made prior to Perelman's announcement.

It could well be that the solution of the Poincaré problem will have consequences for our knowledge of the structure of the universe. Just as Einstein's theory of relativity profited from the improved understanding of geometry worked out by mathematicians in the nineteenth century, so could Poincaré's vision play an important role someday in describing the universe as a whole. However, what will happen remains to be seen. It is known that the universe is locally three-dimensional, and there are theories that the universe is finite but without a boundary. But there is much work to be done before the missing assumption of a simple relationship will be supported both theoretically and experimentally.

---

[2]Translator's note: Instead of publishing his result in a refereed journal, which is a requirement for the prize, Perelman has presented a sketch of how to solve the problem in a series of Internet postings. Since this article appeared, several papers have been published filling in the details. However, the prize has not yet been awarded. Perelman has not said whether he would accept the million-dollar prize if it were offered to him, though he did decline the Fields Medal that was to be awarded to him for this work.

[3]Translator's note: Caution is well advised. After Andrew Wiles announced his proof of the Fermat conjecture in 1993 (cf. Chapter 89) and mathematicians throughout the world began poring over his work, a significant gap was found in the argument, and it took until 1995 for Wiles, with the assistance of colleague Richard Taylor, to fill it in.

Chapter 94

Complex Numbers Are Not So Complex as Their Name Suggests

If one squares any garden-variety number, the result is positive: 3 times 3 is 9, while minus 4 times minus 4 is also positive, namely 16. Therefore, it is difficult to imagine a number whose square is negative.

This issue caused mathematicians considerable grief several centuries ago when they began dealing with the problem of solving polynomial equations. The solution to the problem consisted of an unsurprising part and a spectacular part. It was not at all surprising that a whole new kind of number had to be considered in dealing with equations that could not be solved with the old methods.

The reader may well remember something like this from school. Even after you had progressed to the point of mastering the multiplication table, you were still stymied by the impossibility of finding a number x for which $x + 3 = 1$. The correct solution, namely $x = -2$, was available only after negative numbers had been introduced. And negative numbers have practical significance as well: for doing computations with debit and credit, working with negative temperatures, and in countless other applications.

The case of the complex numbers is similar. Just as one arrived at the negative numbers by considering an equation that otherwise would have had no solution, the complex numbers will arise as a result of equations like $x^2 + 1 = 0$, which can be solved only by a number x whose square is -1.

The spectacular part was this: this new domain of numbers, which contains the square root of every negative number, has the surprising and pleasant property that in it, *all* polynomial equations can be solved, and it will not be necessary to expand into ever more complex number systems as the equations that we wish to solve become more complicated.

All of this was settled in the eighteenth and nineteenth centuries. Since then, mathematicians, engineers, and physicists have used complex numbers with the same ease and confidence that the ordinary person uses numbers like 3 and 12. Complex numbers can be pictured as points in the plane, and in principle, these numbers are no more problematic than the numbers that we work with every day.

So what good are they? Complex numbers are as necessary to mathematicians, engineers, and physicists as negative numbers are to bookkeepers.

However, it is true that it takes a bit of acclimatization before one is fluent with their manipulation, and one certainly doesn't need them in solving the problems of daily existence. It was certainly a marketing disaster calling such numbers "complex" and "imaginary." Such names give them an aura of mystical otherworldliness that they do not deserve. However, anyone who becomes confused by them is in good company. Such irritation with the irrational is well described by Robert Musil in his novel *The Confusions of Young Törless*:

> "You, did you understand that?"
> "What?"
> "The business about imaginary numbers."
> "Yes. That's not so difficult. You just have to realize that the square root of negative one is the unit of calculation."
> "But that's just the point. There is no such thing. . . "

"Quite right; but shouldn't one try in any case to apply the operation of taking the square root to negative numbers as well?"
"But how can you when you know, and know with mathematical precision, that it is impossible?"

What You *Really* Need to Know about Complex Numbers

You will get along just fine with complex numbers if you know the following facts:

(1) **They can be pictured in the plane:** Think of a point in the plane with the usual rectangular coordinates. In Figure 1 you can see depicted the point with coordinates $(2, 3)$.

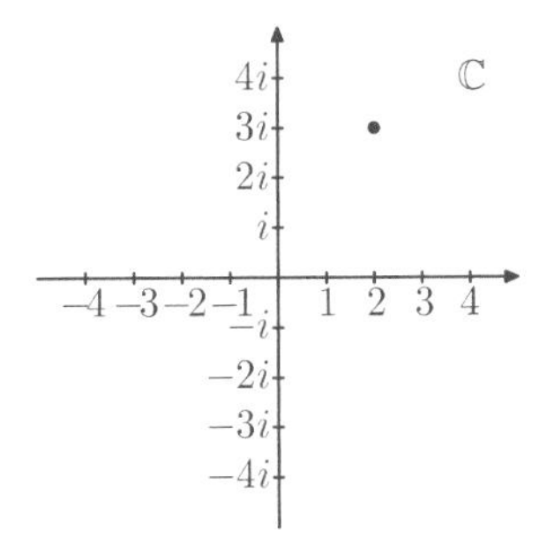

Figure 1. The complex plane.

We shall no longer speak of points in the plane, but of complex numbers. For a point with coordinates (x, y) we write $x + y \cdot i$. For example, the point in the figure is the "number" $2 + 3 \cdot i$. That may seem a bit mysterious, but what is important is that one can write any point such as, for example, $(12, 14)$ as a number (result: $12 + 14 \cdot i$), and conversely, one can determine the unique point associated with every complex number; for example, the number $3 + 2.5 \cdot i$ is associated with the point $(3, 2.5)$.

(2) **It is easy to calculate with complex numbers:** Addition is defined as follows. For example, to add $2 + 3 \cdot i$ and $7 + 15 \cdot i$, you simply add the "real" parts (without the i) and the "imaginary" parts (with the i) separately: Thus one obtains $9 + 18 \cdot i$, since $2 + 7 = 9$ and $3 + 15 = 18$. Similarly, $-9 + 5.5 \cdot i$ is the sum of $-6 + 3 \cdot i$ and $-3 + 2.5 \cdot i$. Further examples should be unnecessary.

For multiplication, you simply multiply as you would a pair of polynomials, with the proviso that whenever you encounter the product $i \cdot i$, you are to replace it with -1. As an example, let us multiply together $3 + 6 \cdot i$ and $4 - 2 \cdot i$. The "usual" calculation yields $12 - 6 \cdot i + 24 \cdot i - 12 \cdot i \cdot i$. Now using the rule that $i \cdot i = -1$, the last term in this expression simplifies to $-12 \cdot (-1) = +12$. Therefore, the product of the two numbers is the complex number

$$12 - 6 \cdot i + 24 \cdot i + 12 = 24 + 18 \cdot i.$$

We note that since multiplying i by itself yields -1, the equation $z^2 = -1$ now has a solution.[1]

(3) **Now all polynomial equations can be solved—linear, quadratic, cubic, etc:** What this means is that regardless of the degree of the polynomial, and regardless of whether the coefficients are real or complex, there will always exist a complex number of the form $z = x + y \cdot i$ that satisfies the equation. Thus, for example, it is guaranteed that there is some complex number that satisfies the equation $z^{10} - 4z^3 + 9.2z - \pi = 0$. The importance of this fact cannot be overstated. For example, when an engineer investigates the frequency characteristics of an electrical circuit or a giant radar antenna, the solutions of an equation in complex numbers will tell whether the system is susceptible to a loss in stability.

[1]Note that $z^2 = z \cdot z$, the usual notation, is used for complex numbers as well as for real numbers.

That there is a solution to an arbitrary polynomial equation is the principal significance of the complex numbers. That this was so was conjectured as early as the seventeenth century, but the first rigorous proof was given by the great Carl Friedrich Gauss in 1799.[2]

[2] Cf. Chapter 25.

Chapter 95

M. C. Escher and Infinity

The work of the Dutch graphic artist Maurits Cornelis Escher is not held in particularly high esteem by connoisseurs of fine art. However, the extent to which such opinion may be justified is not the subject of this chapter. What is indisputable is that Escher's images are interesting from a mathematical point of view on a number of counts, all of a geometric nature.

The first point of geometric interest is *tiling the plane*. The idea of tiling is to cover the entire plane by repeating a basic pattern. Many tiled floors demonstrate the principle. For example, if you draw a square, it can be repeated infinitely in all directions, as though on an infinite chessboard. You can make things more interesting by artful coloring, reflecting, and cutting out a piece on one side and inserting it on the other. Instead of squares, you can use triangles, parallelograms, trapezoids, or hexagons. Indeed, there is quite a variety of ways in which the plane can be tiled.

Escher approached this problem like a mathematician: how many essentially different ways of tiling are there, and how can they be described? Although he was a mathematical amateur, he came up with a complete classification, and all possibilities are realized in his work. Mathematicians working with similar problems were impressed with his results.

What is also impressive is Escher's success at representing infinity. No matter how large the canvas, one can present only a portion of any tiling. Escher found two solutions to this problem. The first is to use a different unbounded surface from the plane; for example, he tiled the surface of a sphere, which resulted in patterns that flow continuously into one another. In his second solution, Escher made use of a mathematical development from the nineteenth century that he learned from the mathematician H. S. M. Coxeter, namely non-Euclidean geometries, which allow one to model infinity on what in the usual geometry is a finite portion of the plane. It is thus that Escher's famous snake and fish motifs arose.

Finally, we should mention Escher's "impossible" pictures, such as the staircase that while finite, spirals forever upward. Each detail looks reasonable enough, but the viewer is unable to put together all the local impressions to form a three-dimensional image that conforms with reality.

Even if we used the terms "local" and "global" in their mathematical sense, there would still be an inexplicable remainder that belongs to the psychology of perception. In looking at Escher's pictures, it becomes clear that the eye generally works as a quiet and unobtrusive censor that sends only preprocessed messages to our consciousness.

Do-It-Yourself Escher

Would you like to make your own Escher print? Perhaps you require a space-filling motif for some wallpaper or gift wrap. There is no limit to the possibilities that await the enterprising hobbyist, though it must be said that it has long been known that there are only twenty-eight essentially different construction methods (all of which were actually used in Escher's images).

What follows is a relatively simple pattern to get you started. In tiling jargon it is *basic type CCC*. To understand it, you will have to know what a *C line* is.[1] It is a curve that joins two points A

[1] The "C" stands for "center."

and B such that the midpoint of the line segment from A to B lies on the curve and such that rotating the curve 180 degrees about the midpoint yields the exact same curve. Some examples are shown in Figure 1.

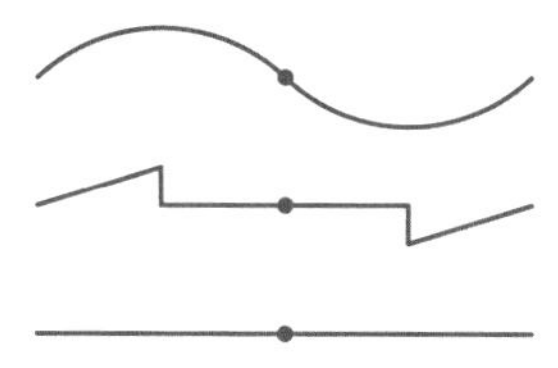

Figure 1. Three C lines.

Now you can let your imagination take flight. Take a clean sheet of paper, draw three points P, Q, R, and draw some C lines: one from P to Q, one from Q to R, and one from R to P. They can be any C lines you like, and in creating the surface—let us call it F—bounded by these three lines, only the degree of your creativity will limit you. Remarkably, one can now seamlessly tile the plane with copies of F. Try it: make a dozen copies of your figure and cut them out. And then you rotate and tile. Instead of trying to explain in detail, here is a picture using the C lines of Figure 1:

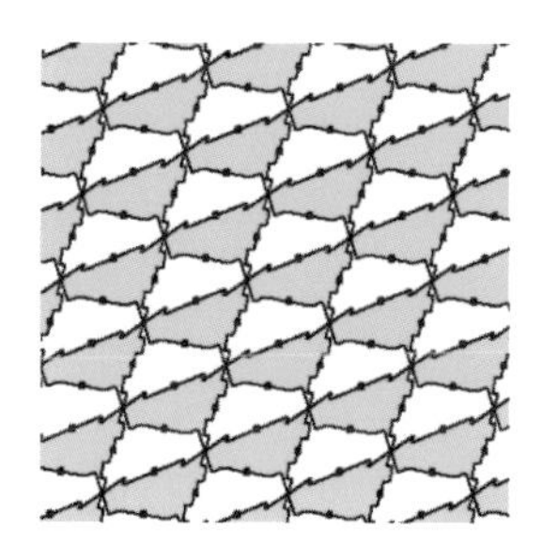

And if you have shaped F in such a way that it represents an angel, devil, fish, bird, or whatever you like, your masterpiece will look just like one of Escher's graphics.

If type CCC is too complicated for your taste, you may prefer type TTTT.[2] Begin with a parallelogram. We are going to name the

[2]T stands for "translation."

vertices A, B, C, D, beginning with the vertex on the lower left and proceeding counterclockwise. Draw a curve of your choosing connecting A and D, and then move a copy of it to the right until it joins B with C. Now draw a curve connecting A with B, and by translating a copy upward, you will have a line connecting D and C. You now have a "mesh" that can be copied left, right, above, below to tile the plane. Because of the construction, these translated building blocks fit seamlessly together. Two examples are presented in Figure 2.

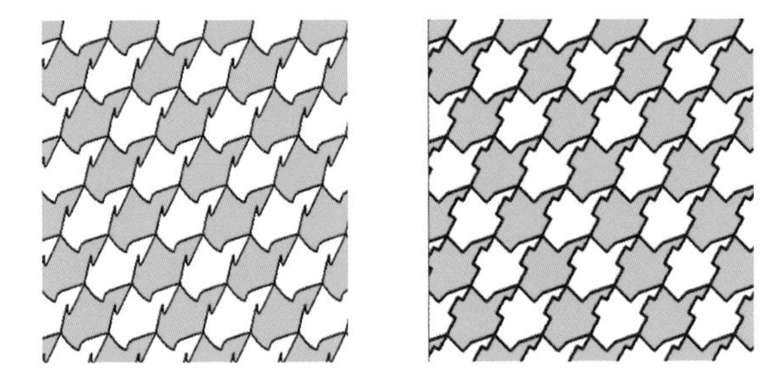

Figure 2. Examples of plane tiling of type TTTT.

Chapter 96

A One at the Beginning Is Much More Likely Than a Two

Did you ever notice? When you look at tables of numbers, you see that the fundamental law of equality is violated: you would assume that the first digit that appears should be 1, 2, and so on with equal likelihood. However, such is not generally the case. This is an example of Benford's law, named for the physicist Frank Benford (1883–1948). One should not take the word "law" too seriously; it has nothing of the inevitability of Newton's laws of celestial mechanics, according to which nature can be described by statutes to which it rigorously adheres. Rather, Benford's law is an attempt at a qualitative explanation.

We are dealing with a variation on the theme "chance covers its tracks." As a first attempt at explanation, imagine, if you would, a simple dice game. The "game board" consists of adjoining fields labeled 0, 1, 2, etc. You begin at 0, and then you proceed as many steps as are shown by the roll of a die. If the first rolls were 1, 6, 2, then you would move to square 1 on your first turn, then to square 7 ($= 1 + 6$), and then to square 9 ($= 7 + 2$).

There is then no way of determining whether square 101 will be landed on with greater probability than square 201 on a particular move of the game, since "far out," all squares have an equal chance. Even if the die is not fair, after a few moves, one has no idea where one will end up.

And now back to Benford's law. In the dice game described above, we were dealing with additive random influences. As a rule, however, the factors that influence a quantity are multiplicative. (For example, double the precipitation results in double the amount of available water for irrigation. The "2" must be multiplied, not added.) A technical trick converts multiplication to addition: by passing to logarithms, multiplicative problems are turned into additive ones. This leads to the result that the logarithms of quantities whose appearance depends multiplicatively on random influences must be uniformly distributed. For the quantities themselves, this means that the digit 1 must appear more frequently at the beginning than 2, which is more frequent than 3, etc.

Why is this? Let us consider two-digit numbers. The base-10 logarithms between 1 and 2 represent the numbers between 10 and 100. (The logarithm of 10 is 1, and the logarithm of 100 is 2.) However, the logarithm of 20 is about 1.3, and so about thirty percent of the logarithms between 1 and 2 (from 1.0 to 1.3) represent the numbers from 10 to 19. The logarithm of 30 is about 1.5, so the two-digit numbers beginning with 2 are represented by only about 20% ($1.5 - 1.3 = 0.2$) of the logarithms. By the time we get to the 90's, we can calculate that the logarithm of 90 is about 1.95, and so the two-digit numbers beginning with 9 are represented by only $2.0 - 1.95 = 0.05 = 5\%$ of the logarithms.

Are you skeptical? Think of a four-digit number and write a "1" in front of it. Then let Google search for this (now a five-digit) number. Do the same with 2, and so on. The number of hits goes successively downward, which should convince any doubters.

A Google Experiment

The source of Benford's law can be seen clearly by viewing a table of logarithms in your local library. One used to use such tables as a way to carry out complex multiplications.[1] Benford noticed that the pages belonging to the smaller initial numbers are more worn than those with larger numbers. Benford then tackled the problem systematically and arrived at his law. It is, as has been mentioned, not a real "law," and our attempt here is only one of many possible ways of trying to explain it.

But there can be no doubt that it really exists. Here is a Google test from December 2005 with the randomly selected sequence 3972 and the numbers 1 through 9 prefixed to it. The number of hits as well as the theoretical (percentage and actual) values are given:

Search	**Actual Hits**	**Theoretical (%)**	**Theoretical Value**
13972	389,000	30.1	346,000
23972	232,000	17.6	203,000
33972	136,000	12.5	144,000
43972	117,000	9.7	112,000
53972	71,400	7.9	91,000
63972	65,300	6.7	77,000
73972	44,600	5.8	68,000
83972	54,100	5.1	59,000
93972	42,300	4.6	53,000

The actual values are somewhat higher at the beginning of the table and somewhat lower at the end than what is predicted by Benford's law. However, the observed result can be seen as a good qualitative verification of the theory.

[1]See also Chapter 36.

Chapter 97

The Leipzig Town Hall and the Sunflower

The golden ratio is one of the most important numbers in all of mathematics. Recall that one can construct a rectangle in which the ratio of the longer side to the shorter side is equal to the ratio of the sum of the two sides to the longer side. This ratio is called the *golden ratio*. To determine this ratio exactly, let us denote the length of the longer side by x and let the smaller side have length 1. Then by our requirement, we must have

$$\frac{x}{1} = \frac{1+x}{x}.$$

Multiplying the equation through by x and rearranging terms leads to the quadratic equation

$$x \cdot x - x - 1 = 0,$$

and the quadratic formula then yields two solutions,

$$x = \frac{1 \pm \sqrt{5}}{2}.$$

Only one of these solutions is positive, and it is equal to 1.6180....

Some are of the opinion that rectangles whose sides are in the golden ratio are particularly aesthetically pleasing. Indeed, the golden ratio is to be found frequently in architecture, for example in the outlines of ancient Greek temples and in modern buildings as well (for instance, the tower on the Leipzig town hall is divided into sections that correspond to the golden ratio). However, in our daily lives, where we are confronted with pieces of paper in the international standard A series, for which a sheet folded in half has the same proportions as the original, we have become more accustomed to the ratio of the sides being equal to the square root of 2, or 1.414....

The importance of the golden ratio is revealed in the fact that it pops up in one way or another in almost every branch of mathematics. It is certainly no surprise that it plays a role in geometry, since the term was introduced as a result of a geometric problem. But it also appears in situations in which only numbers are involved. For example, consider the famous *Fibonacci sequence*. It begins $1, 1$, and then each succeeding term is the sum of the two that precede it. Therefore, after $1, 1$, we have $2, 3, 5, 8, 13, 21$, and so on. The quotient of two successive terms approaches ever more closely the golden ratio. Even $21/13 = 1.615\ldots$ is a good approximation.

Fibonacci numbers pop up in nature, for example, in the arrangement of the seeds of a sunflower. And if you have a tape measure handy, you can search for the golden ratio on your own person. The relationship "distance from elbow to fingertips divided by distance from elbow to wrist" is just one of many possible examples.

However, such results lead one quickly into the realm of speculation. Perhaps the ratio of the number of "bad" characters in Grimm's fairy tales to the number of "good" characters is equal to the golden ratio.

Continued Fractions

The golden ratio is a remarkable number for yet another reason. This time it comes about in a type of *approximation*. In dealing with numbers that cannot be represented as fractions, it is convenient to replace such a number with a fraction that has a relatively small numerator and denominator and that represents a good approximation to the original number. For example, the fraction $22/7 = 3.14285\ldots$ represents a good approximation to π, whose value to five decimal places is 3.14159. This approximation was known to the Egyptians 2,500 years ago, and such an approximation suffices for many everyday applications.

The best rational approximations to a number are obtained by *continued fractions*. These are fractions that arise from a rather complicated process. This is how it works.

A continued fraction is written as a finite sequence of natural numbers enclosed in square brackets. The notation is interpreted as follows:

$$[a_0] = a_0,$$

$$[a_0, a_1] = a_0 + \frac{1}{a_1},$$

$$[a_0, a_1, a_2] = a_0 + \frac{1}{a_1 + \frac{1}{a_2}},$$

$$[a_0, a_1, a_2, a_3] = a_0 + \frac{1}{a_1 + \frac{1}{a_2 + \frac{1}{a_3}}},$$

$$[a_0, a_1, a_2, a_4] = a_0 + \frac{1}{a_1 + \frac{1}{a_2 + \frac{1}{a_3 + \frac{1}{a_4}}}}.$$

If that seems too abstract, here are a couple of concrete examples:

$$[3, 9] = 3 + \frac{1}{9} = \frac{28}{9},$$

$$[2, 3, 5, 7] = 2 + \frac{1}{3 + \frac{1}{5 + \frac{1}{7}}} = \frac{266}{115}.$$

If one approximates a number by the best-possible continued fraction, then the larger the numbers in the continued fraction, the better the approximation in general. That is because all the numbers after the first appear in the denominator of the fraction that the continued fraction represents, and the larger the numbers, the faster the denominator grows; therefore, more decimal places of the number being approximated will be accurately represented.

Returning to the golden section, this number has the remarkable property that among all irrational numbers, it is the one that is least well approximated by a continued fraction, in the sense that the numbers in its representation are as small as possible and thus it takes a relatively large number of terms to approximate the number to a given accuracy. Indeed, the best continued fractions for the golden section are $[1]$, $[1,1]$, $[1,1,1]$, $[1,1,1,1]$, and so on. This fact plays an important role in *KAM theory*.[1] One can conclude from this theory that a vibrating system whose frequency relation is the golden ratio is particularly insensitive to perturbations.

A Riddle

There is a riddle well known to Internet surfers that has an indirect relationship to the Fibonacci numbers. Figure 1 shows a triangle that has been divided into four regions. After rearranging the segments, the same triangle appears, but now a small piece is missing.

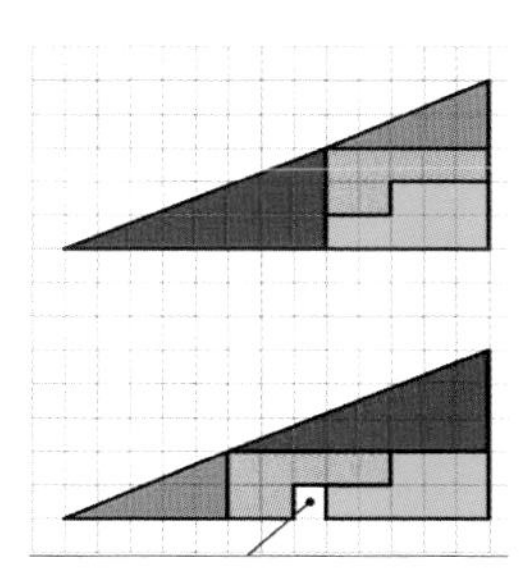

Figure 1. What happened to the missing piece?

[1] KAM refers to the names of the three mathematicians who developed it: Andreĭ Kolmogorov, Vladimir Arnold, and Jürgen Moser.

The solution to the riddle will be given at the end of this article.

The Pacioli Icosahedron

The Italian mathematician Luca Pacioli (1445–1517) discovered an interesting relationship between the golden ratio and the Platonic solids.

Take three rectangles with sides whose lengths correspond to the golden ratio: the longer side is approximately 1.618 times the shorter side.

These rectangles are then interpenetrated perpendicularly, as depicted in Figure 2. If they were made out of wood, you would have to do a bit of sawing. And now for the surprise: if you join the vertices of the rectangles, with string for example, you get a Platonic solid, an icosahedron.

Figure 2. Pacioli's icosahedron.

As with the "most beautiful formula" of Chapter 55, this is an impressive example of how surprising relationships are to be found between diverse branches of mathematics.

The Solution to the Triangle Puzzle

The reason that a little chunk of space can appear or disappear when the pieces are reassembled is that the figure isn't really a triangle at all. In the top picture of Figure 1, what looks like a triangle really has the "hypotenuse" bent inward a bit, while in the lower image, it bulges out a little.

You can convince yourself of this: the red triangle rises at a rate of $\frac{3}{8} = 0.375$, while that of the green triangle is $\frac{2}{5} = 0.4$. The numbers appearing in these two fractions, $2, 3, 5, 8$, belong to the Fibonacci sequence, and the fact that $\frac{2}{5}$ is close to $\frac{3}{8}$ is related to the convergence of these fractions drawn from the Fibonacci sequence to the golden ratio.

Chapter 98

Information Optimally Packaged

The necessity of transmitting information among people with a common interest appears in the most varied areas of life. How does a group of jazz musicians understand among themselves the harmonies over which they are improvising? How does a couple manage to dance the tango? What universal product code belongs to the article that one would like to pay for at the checkout counter? In mathematics, such questions have led to a new field called coding theory. It deals with the "optimal" packaging of information, where "optimal" is to be construed according to the situation.

For example, how can a message be sent so that it can be read by the recipient even if some garbling of the text occurs during transmission? A naive solution to the problem might be to send the same message a number of times, say five. It is then extremely unlikely that the message would be defective or entirely illegible at exactly the same place in all five transmissions, and so the recipient would be able to piece together a valid text from the five copies.

The great disadvantage of such a method is that it is extremely uneconomical, and indeed, more-elegant solutions to the problem are available. One approach is—using computers, of course—to convert the message letter by letter into a string of ones and zeros. Let us suppose that each symbol is encoded by a particular sequence of length ten. As a simple test of whether the transmission arrived intact, we could append a one or a zero to the code depending on whether the number of ones in the ten-bit encoding is even or odd. The recipient then knows that something has gone awry if the "check digit" doesn't correspond to the transmitted code. The accuracy of the transmission can thus be checked with only a ten percent increase in complexity. Of course, in the case of an error, it is not at all clear where the error—if indeed the transmission error occurred in the code and not in the check digit—lies and how it is to be corrected. But there are more-advanced coding techniques that take care of these issues.

The high art of coding theory is to send messages reliably even over very noisy communication channels. The received message can truly be relied on as correct. A typical example of successful communication is the images sent from distant space probes—from Mars, for example. In a less-spectacular area, coding theory is gainfully employed in your CD player. It is only because of the built-in coding theory that your CD player can play your favorite song perfectly, even if the disc has accidentally become badly scratched.

Error-Correcting Codes

A check digit, as described above, allows one to tell whether a single transmission error occurred somewhere. If you received the message 011000001011, you would know that something was amiss, since every packet of eleven bits should have an even number of ones.

Such simple tests are often sufficient, at the supermarket checkout, for example, where if the apparatus detects an error in reading a universal product code, it fails to register the item, and the cashier has simply to scan the item again.

However, sometimes it is important to know just which bit was incorrectly transmitted. Then the error can be corrected and the original transmittal reconstructed.

The first easily implemented such code was described in 1948 by R. W. Hamming. At that time, the idea of a self-correcting code was revolutionary.

To describe Hamming's idea, consider a sequence of zeros and ones of length four. We shall denote the general sequence by $a_1a_2a_3a_4$. Thus for the sequence 0110, we have $a_1 = 0$, $a_2 = 1$, $a_3 = 1$, $a_4 = 0$. We are going to add *three* check bits, which we will call a_5, a_6, a_7. They are determined by the following rules:

- If the number of ones among a_1, a_2, a_4 is odd, then set $a_5 = 1$; otherwise, set $a_5 = 0$.
- If the number of ones among a_1, a_3, a_4 is odd, then set $a_6 = 1$; otherwise, set $a_6 = 0$.
- If the number of ones among a_2, a_3, a_4 is odd, then set $a_7 = 1$; otherwise, set $a_7 = 0$.

Then the sequence of seven symbols is sent: $a_1a_2a_3a_4a_5a_6a_7$. In our example, in which the sequence 0110 is to be sent, the sequence is extended to 0110110. For example, the last 0 arose because among the numbers a_2, a_3, a_4 (thus the numbers $1, 1, 0$), there is an even number of ones.

And what's the benefit? Suppose that an error occurred during transmission, causing one of the bits to be switched. It is known that in each of the sequences $a_1a_2a_4a_5$, $a_1a_3a_4a_6$, and $a_2a_3a_4a_7$ there should be an even number of ones. If, however, bit a_1 were received incorrectly, then $a_1a_2a_4a_5$ and $a_1a_3a_4a_6$ would have an odd number of ones, while $a_2a_3a_4a_7$ would be fine. That is, the pattern for a switched a_1 bit would be odd, odd, even. And what about the other bits?

- a_2 switched: odd, even, odd.
- a_3 switched: even, odd, odd.
- a_4 switched: odd, odd, odd.
- a_5 switched: odd, even, even.

- a_6 switched: even, odd, even.
- a_7 switched: even, even, even.

From this analysis, by knowing in which of $a_1a_2a_4a_5$, $a_1a_3a_4a_6$, and $a_2a_3a_4a_7$ the incorrect number of ones occurs, one can determine precisely which bit is incorrect, correct it, and thereby restore the original transmission.

> An example: Suppose the sequence 0110110 was garbled by having its first bit switched, resulting in 1110110. We consider the number of ones in $a_1a_2a_4a_5$, $a_1a_3a_4a_6$, and $a_2a_3a_4a_7$, that is, in 1101, 1101, and 1100. The pattern is odd, odd, even, and so we may conclude that the error was in the first bit.

Note that this method allows us to determine not only whether a message bit was switched, but whether the defective bit was one of the check bits. If such is the case, we know that the message is intact but can conclude that the communication channel is noisy.

Note also that this Hamming code does not allow us to determine whether more than one error is present. However, more-refined error-correcting codes can deal with two, three, or more potential errors in zero-one sequences of arbitrary length, detecting and correcting them. As we have mentioned, such error-correction ability is of fundamental importance in compact disc players, since creating discs that were one hundred percent perfect would be prohibitively expensive.

Chapter 99

Four Colors Suffice!

Take a sheet of paper and draw a map of an imaginary continent, sketching in the various countries. Now take out your watercolors and color the countries, making sure that no two countries sharing a common border have the same color. You will soon discover that four colors always suffice for the map coloring, even if for a complicated map you may have to use a bit of ingenuity. An example of a map requiring four colors is shown in Figure 1.

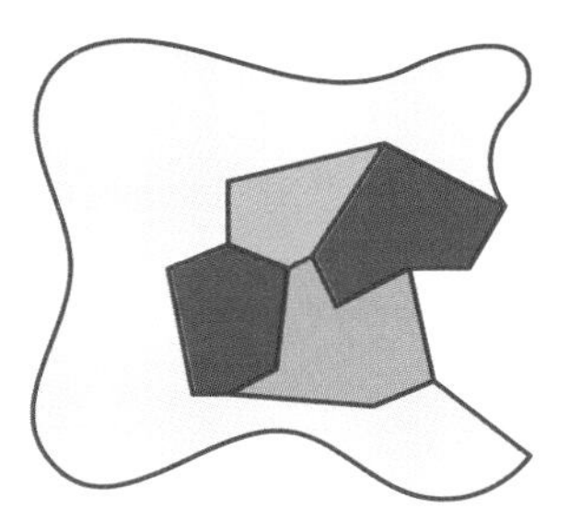

Figure 1. Four colors are necessary for this map.

This fact was observed in the nineteenth century. However, mathematicians are never content with a large quantity of experimental observation. A proof was sought that guaranteed that four colors *always* suffice, no matter how many countries and no matter how intertwined they may be.

Many attempts at a proof were made, but the problem turned out to be a tough nut to crack. The mathematical world had to wait until 1976 until the four color theorem was finally proved by Kenneth Appel and Wolfgang Haken, at the University of Illinois: "Four colors suffice!" read the postmark from the University of Illinois Department of Mathematics.

There was one small unsettling feature about the proof, which has kept the discussion of the issue alive. No one doubts that the proof is correct. However, important parts of the argument rely on massive computer calculations that no army of human calculators could ever hope to check. For mathematicians, who have relied on paper and pencil for centuries, this was a new and unsatisfactory situation, and to this day, many have not yet come to terms with this phenomenon. Even if dozens of computers confirm the validity of the argument, that can never give the same satisfaction as an individual's "personal" verification of the steps leading to a proof.

For those interested in practical applications, the four color theorem is not of great interest. It belongs in the category of such theorems as "there are infinitely many prime numbers" and "the number π is transcendental." The fascination is rather that a problem so easily stated, that defied proof for so long, has finally been settled. Forever. For any number of countries. And sooner or later, someone will come up with a proof that does not rely on computers, and then everyone will be satisfied.

Maps and Graphs

The four color problem provides a nice example of how mathematicians reduce problems to their essentials. In choosing the colors, it is completely irrelevant what the boundaries of the countries look like. One cares only whether two countries share a common border. Therefore, the map-coloring problem is transformed into a graph-coloring problem:

> Represent each country by a point on a piece of paper. If two countries share a border, draw a line between the corresponding points.

Such a system of points and lines is called a *graph*.[1] Graphs of this type are to be found outside of mathematics as well, such as in the form of subway maps and computer flowcharts.

The map of the sixteen German states can be converted to a graph as shown in Figure 2.

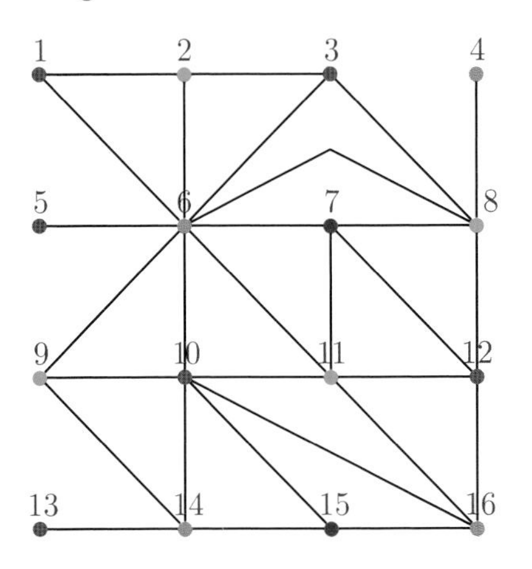

Figure 2. The graph of the German states.

The numbers in the figure represent the following German states:

1: Hamburg	2: Schleswig-Holstein
3: Mecklenburg-Vorpommern	4: Berlin
5: Bremen	6: Niedersachsen
7: Sachsen-Anhalt	8: Brandenburg
9: Nordrhein-Westfalen	10: Hessen
11: Thüringen	12: Sachsen
13: Saarland	14: Rheinland-Pfalz
15: Baden-Württemburg	16: Bayern

Our coloring problem takes the following form: can one color the sixteen points with four colors in such a way that no two points that are connected by a line have the same color?

For the German states, such a coloring is shown in the figure.

[1]This definition of "graph" is completely independent of the graphs of functions that you drew in school.

The Farmer, the Goat, the Wolf, and the Cabbage

As a very simple example of how a suitable representation as a graph can make a problem easy to solve, we recall the following classic riddle:

> A farmer has to transport a goat, a cabbage, and a wolf[2] across a river in his rowboat. Unfortunately, the boat is small, and the farmer can take only one "passenger" at a time with him. Complicating the problem is the fact that there are certain pairs that cannot be left alone on the shore (the wolf would eat the goat; the goat would eat the cabbage).

How should the transportation be organized? The solution is clear if one reformulates the problem suitably. We will call the banks of the river the left bank and the right bank, and suppose that at present, all the players are located on the left bank. How do we proceed so that the wolf doesn't eat the goat, and the goat the cabbage? We represent each possible scenario by a point labeled with the initials of the passengers that are currently on the left bank. See Figure 3, which shows the ten possible scenarios.

The symbol ∅ represents the empty set (as in Chapter 12), indicating that all the passengers have been transported to the right bank. The five points on the left-hand side of the graph represent scenarios in which the farmer is on the left bank, while for those on the right-hand side, the farmer is on the right bank.

For example, the second point from the top in the left-hand column, labeled G, represents the scenario in which goat and farmer are on the left bank (and therefore with wolf and cabbage on the right bank).

Points representing prohibited scenarios are not indicated. For example, GWC on the right would indicate the farmer on the right and all the passengers on the left, in which case perhaps first the goat would eat the cabbage and then the wolf the goat. That is why GWC on the right does not appear in the graph.

[2] In some variants it is a ferocious dog.

Now let us look at the permissible transports. We join any two scenarios with a line if the farmer can get from one state to the other by transporting a single passenger. For example, the line running from GWC on the left to WC on the right indicates that the farmer rows the goat from left to right, leaving the wolf and cabbage behind. Notice that none of the other states on the right can be reached from GWC, since that would involve transporting more than one passenger, which is forbidden by the terms of the problem.

In the language of graphs, the problem reads thus: is there a path over permissible paths from the upper left-hand point (everybody on the left bank) to the upper right-hand point (everybody on the right bank)? That such is indeed the case, and the actual series of transports that must be made, can be read off at once from the graph.

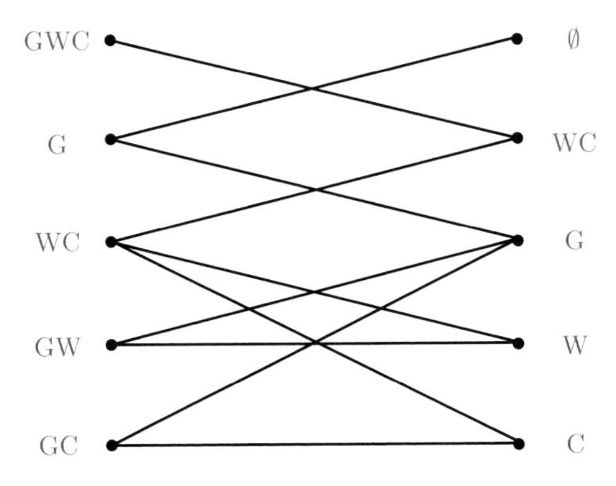

Figure 3. The graph for the transport problem.

Chapter 100

Mathematics Makes Billionaires

When Google went public, the founders Sergey Brin and Lawrence Page suddenly joined the ranks of the world's wealthiest men.

Anyone wishing to imitate their success would first have to get hold of a gigantic computer and then create a catalog of all the world's web sites. There are currently about twenty billion web pages.[1] For each page, one has to create an index of all terms of interest. That is surely a time-consuming task, but for a team of talented programmers, it is no insuperable challenge. Of course, all the really hard work is left to the computer!

Suppose that all these tasks have been satisfactorily completed. Unfortunately for you, you are still in no position to offer a competitive search engine. The reason is the enormous size of the Internet. Given a particular query, such as all sites containing both "USA" and "hurricane," it is not difficult to produce all the web pages with the two search terms. The problem is how to present the results, since typically a search returns from hundreds of thousands to millions of hits. No one has the time or patience to sift through all those pages. One would like to have the most important pages presented first.

[1] To get an idea of the size of this number, note that the distance along the Earth's surface from the North Pole to the South Pole is about 20 billion millimeters.

Those who have done much "googling" know that Google has done a surprisingly good job of solving this problem, for generally one finds what one is looking for among the first dozen or so pages.

The secret is in the correct definition of "important." Google's great idea was to consider a web page important if many important web pages refer to it. If one thinks of a web page as a point, with two points connected by an arrow if the page at the arrow's tail has a link to the page at the arrow's head, then the Internet can be pictured as a network of about two billion points connected by many more billions of arrows. A tiny segment of this network is illustrated in Figure 1.

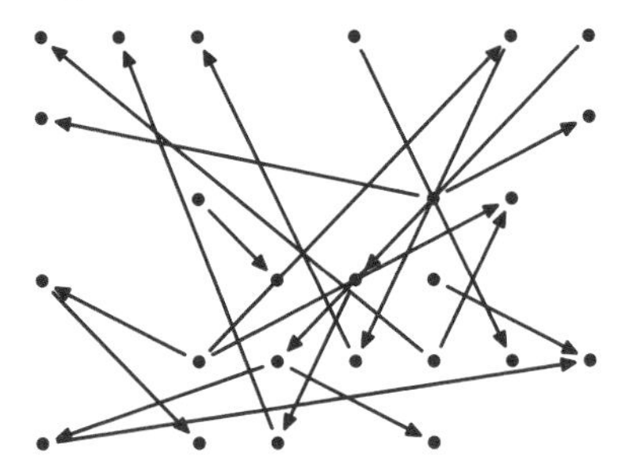

Figure 1. A network of web pages pointing to other web pages.

Pages at which many arrows terminate are "important," particularly if the arrows emanate from "important" web pages. If we number the web pages $1, 2, \ldots$ and assign a weight W_i to page i as a measure of its importance, there should be a relationship among these numbers.

For example, if page 2 is linked to from page 5, and page 5 is linked to by three pages altogether, then page 2 "inherits" one-third of the importance of page 5. Perhaps page 7 also links to page 2 and has ten outgoing links altogether. Then page 2 inherits one-tenth of W_7. Suppose that is all: no other pages link to page 2. That leads to the equation

$$W_2 = \frac{W_5}{3} + \frac{W_7}{10}.$$

Most web sites are connected in complex ways, and altogether, we end up with twenty billion equations in the twenty billion unknowns $W_1, W_2, \ldots$.

The math you learned in school is not going to be of much help here. For most readers, two equations in two unknowns is the most they ever had to deal with. But even for professional mathematicians this number is too large to be handled by the standard methods, even given that some optimization problems with several hundred thousand or even several million variables can be solved.

Another path leads to the goal: one takes a "random walk." Imagine an Internet addict, let us call her Isabella, who starts at, say, the web page www.ams.org. Isabella then searches the page at random for a link to another page and clicks on it. Arriving at that page, she sees another set of links, and randomly clicks on one. Thus a random walk through the worldwide web has begun, in which out of logical necessity, the "more important" pages are visited more frequently than the "less important" ones. It is a remarkable fact that the relative frequencies of the visits satisfy the system of equations described above. In short, the importance of a page can be determined by measuring the percentage of time that a surfer is to be found there.

However, doing these calculations doesn't seem any easier than solving the initial problem. And if an exact solution is required, that is indeed the case. However, an approximate solution (in which, say, you would be satisfied with five decimal places of accuracy) can be had in several hours.

Now the search engine is fully functional, since once you have the levels of importance, the rest falls into place: simply search for all pages with "USA" and "hurricane" and return them in order of importance.

That is a first approximation to Google's methodology. The details and refinements are complex and as secret as the formula for Coca Cola. As an example of a necessary refinement, we might mention that our random surfer Isabella would have a problem if she arrived at a site without any further links. To avoid such a scenario, at each step in the search, the links on a page are ignored with a certain probability p and one goes instead to a randomly selected web page somewhere in the Internet. (That may not be easy for Isabella, but with a directory of all web pages it is no problem.) It is rumored that Google uses the probability $p = 0.15$, which seems to be a good value, based on experience. Google is also constantly working to minimize the effects of "Google bombing," which is a strategy of puffing up the importance of one's web site by artificially increasing the number of links to one's own pages.

Google's competitors have also not been asleep at the switch. There is constant work underway at developing new approaches and calculational techniques for evaluating the "importance" of web sites. For different users and different combinations of search terms, "important" can mean something quite different. But adding in such complexity then leads to the problem of whether—as with Google—the results of a search can be returned within a fraction of a second.

Further Reading

In recent years, a number of popular books on mathematics have appeared. Some of these have been mentioned in this book. Here is a small selection of additional books that may be of interest.

M. Aigner and G. Ziegler.
Proofs from the Book. Springer, 2003.
This book collects some of the most beautiful proofs in mathematics. Recommended to all who have some background in mathematics.

A. Beutelspacher.
Cryptology. Mathematical Association of America, 1996.
For all who wish to learn more about cryptography.

J. Bewersdorff.
Luck, Logic, and White Lies: The Mathematics of Games. A. K. Peters, 2004.
Several chapters of *Five-Minute Mathematics* are devoted to games of chance and gambling. This book examines the mathematics of games from chess to Monopoly.

A. Doxiadis.
Uncle Petros and Goldbach's Conjecture: A Novel of Mathematical Obsession. Bloomsbury, 2001.
This wonderfully written novel is about a man obsessed with the solution to Goldbach's conjecture. There is no better description of how someone can be changed by succumbing to the fascination of a mathematical problem.

U. Dudley.
Numerology: Or, What Pythagoras Wrought. Mathematical Association of America, 1997.
A must for those interested in number mysticism.

R. Kanigel.
The Man Who Knew Infinity: A Life of the Genius Ramanujan. Washington Square Press, 1991.
An extensive biography of the mathematical genius Ramanujan.

R. Kaplan.
The Nothing That Is: A Natural History of Zero. Oxford, 2000.
Anyone wishing to know more about zero should read this book.

P. Ribenboim.
The Little Book of Bigger Primes. Springer, 2004.
Five-Minute Mathematics has much to say about large prime numbers. In this book the subject is systematically developed.

M. du Sautoy.
The Music of the Primes: Searching to Solve the Greatest Mystery in Mathematics. Harper Perennial, 2004.
Yet more about primes!

S. Singh.
Fermat's Enigma: The Epic Quest to Solve the World's Greatest Mathematical Problem. Anchor, 1998.
This book is one of the best popular books on mathematics. The reader learns much about mathematics in general and about mathematicians in a presentation of the Fermat problem, which was finally solved by Andrew Wiles.

S. Singh.
The Code Book: The Science of Secrecy from Ancient Egypt to Quantum Cryptography. Anchor, 2000.
Just as full of information and detailed as his book on Fermat's last theorem, in this book Singh discusses the history of cryptography from ancient times to the present day. Needless to say, the RSA method is described in full. However, one learns as well how the Enigma code operated and how its secrets were cracked by the British.

R. Taschner.
Numbers at Work: A Cultural Perspective. A. K. Peters, 2007.
Many of the topics presented in *Five-Minute Mathematics* can also be found in Taschner's book. The author has translated his extensive knowledge

into a book that is exciting to read. The major themes are number and symbol, number and music, number and time, number and space, number and politics, number and matter, number and spirit.

Index